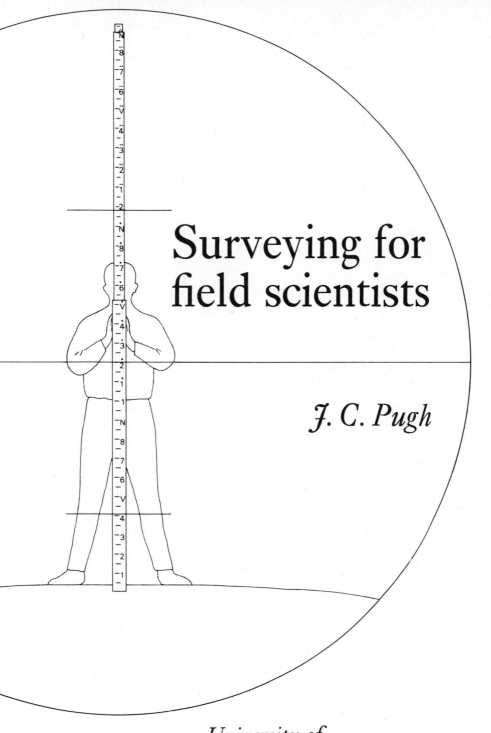

Surveying for
field scientists

J. C. Pugh

University of
Pittsburgh Press

Published 1975 in the U.S.A.
by the University of Pittsburgh Press

First published 1975 by Methuen & Co Ltd
11 New Fetter Lane, London EC4

© 1975 J. C. Pugh

Typeset in IBM by William Clowes & Sons Limited, London,
Colchester and Beccles

Printed in Great Britain by Butler and Tanner Ltd, London and Frome

Library of Congress Catalog Card Number 74-32566
ISBN 0 8229 1120 5

Contents

Preface   ix

List of figures   xi

Introduction   xv
   General principles   xvii
   The structure of the book   xviii
   Typical techniques to be used   xix

1 Provision of fixed marks   1

   Orientation   1
   Sun azimuths   2
   Other methods of orientation   14
   Small triangulation schemes   14
   Station location and signals   15
   Base line   17
   Triangulation observation   22
   Adjustment of errors   24
   Satellite stations   38
   Heights   41
   Trigonometrical observations   41
   Heights by levelling   43

2 The fixation of minor points 47

Theodolite traverses 47
Measurement of distances 49
Measurement of angles 57
Calculation of traverse 59
Traverse heights 66
Mapping of detail with the theodolite 67
Addition of detail to standard traverses 70
Direct-reading tacheometric instruments 73
Angular measurement with the sextant 76
Vertical angles 77
Horizontal angles 78
Plane table traverses 80
Large scale 80
Small scale 84
Plane table traverses with tacheometric measurement 85
Direct-reading tacheometric instruments 86

3 Infilling of further detail 89

Plane table resections 89
Partial resection 89
Full resection. Method 1 (Bessel's) 90
Full resection. Method II (tracing paper) 91
Full resection. Method III (triangle of error) 91
General comments 93
Contouring 94
Theodolite in conjunction with plane table 96
Prismatic compass traverse 97
Traverse adjustment 98
Heights in conjunction with compass traverses 103
The Brunton compass and inclinometer 107
Use as a compass 107
Use as an inclinometer 110
Use of the aneroid barometer 111

4 Instruments 118

General principles 118
The level 119
Levelling staves 122
Adjustment of the level 123
Method of booking 125
The theodolite 128
The tripod 131

The centring device 131
The levelling head 133
The lower plate system 134
The upper plate system 136
The vertical reading frame 137
The telescope and vertical circle 138
Reading devices 139
Adjustments of the theodolite 142
The sextant 145
Tilt correction 150
The Abney clinometer 153
Adjustment 155
The aneroid barometer 156
The plane table 158
Accessories with the plane table 159
The prismatic compass 163
Bearings 166
Battery-operated calculators 166
Laboratory work 166
Field work 167
Radio telephones 167
Control networks 167
Detail fixation 167
Aneroid readings 168
Specialised readings 168

5 **Simple astronomy** 169

Declination and Right Ascension 169
Declination and Hour Angle 171
Altitude and azimuth 173
The astronomical triangle 173
Time 175
Solar time 175
Latitude and longitude observations 176
Position lines 179

Appendices

1 **The geometry of resection** 189

Trigonometrical resection (Collins Point) 189
Plane table (Bessel's method) 191

2 Basic trigonometric functions   194

Topographical survey tables   199

1   Sun's parallax in altitude   201
2   Astronomical refraction   201
3   Corrections for slope for measured distances of 100 units   202
4   Tacheometric tables:
    A. True horizontal distances from stadia intercepts   203
    B. True vertical differences from stadia intercepts   208
5   Natural tangents   213
6   Checking table for traverses   214
7   Tilt correction for horizontal angles measured with the sextant   218
8   Conversion constants   228

Preface

Textbooks on surveying are usually written for professional surveyors. Quite rightly, they stress the need for complete accuracy and they strive for elimination of all possible sources of error. This book seeks no more than to set out methods which a non-specialist may use to produce a map on which errors are not plottable. The techniques suggested are sometimes not professional techniques; refined instrumental adjustments and corrections for greater accuracy are omitted if they will not affect map plotting; the methods of calculation involve logarithmic techniques because the lone worker will not always have access to computer facilities which reduce and simplify his work. Throughout, the idea of a plane table map has been kept in mind as the desirable end-product, and the techniques and tables are intended to simplify such mapping. The tables given in the book are intended for use in the field and have been employed there. If nothing else, it should at least be possible for a complete beginner, with one assistant to carry a staff, to make a reasonably accurate map of an area of half a kilometre diameter, showing detail both of location and altitude.

List of figures

1   Relative positions of sun and R.O. in the calculation of azimuth
2   Intersection of the sun's image in the theodolite telescope
3   Sun observations: error introduced by mean of FL and FR readings
4   Method of booking observations for sun azimuths
5   Method of calculating sun azimuths
6   Correction of observed angles for dislevelment of bubble
7   Atmospheric refraction
8   Solar parallax
9   Relationship of azimuths to sun and R.O.
10 A triangulation control net (Iping Common)
11 Base measurement: inclusion of residuals
12 Base measurement: booking of observations
13 Base extension net
14 Triangulation observations; method of booking; analysis of angles
15 Triangulation net: simple triangles
16 Calculation of differences in Eastings and Northings
17 Triangulation net: improvement with braced quadrilaterals
18 Braced quadrilateral: angular conditions
19 Braced quadrilateral: side conditions
20 Triangulation net: centre-point figures
21 Satellite station and observed triangle
22 Satellite station: site plan of rays
23 Satellite station: local triangle for distance measurement
24 Trigonometrical sights: effect of earth curvature

25   Trigonometrical sights: graph of curvature and refraction effects
26   Levelling: equal lengths for foresight and backsight
27   Levelling: imbalance of errors if lengths not equal
28   Levelling: stadia measurement of distance
29   Control net: insertion of traverses between triangulation points
30   Ranging rod in miniature support tripod
31   Surveyor's chain: distinguishing tags
32   Traverse measurement: method of booking
33   Traverse measurement: necessity of slope corrections
34   Theodolite traverse: method of booking angles
35   Theodolite traverse: angular error due to non-vertical ranging rod
36   Theodolite traverse: method of booking angles at junction point
37   Theodolite traverse: calculation and adjustment
38   Theodolite traverse: relationship of bearings and observed angles
39   Theodolite traverse: relationship of bearings and co-ordinates
40   Theodolite traverse: probable location of gross error
41   Tacheometric measurement of distance on slopes
42   Tacheometric measurement: 10-minute estimation of theodolite readings
43   Tacheometric measurement: method of booking detail in a theodolite traverse
44   Ewing stadi-altimeter: eyepiece images
45   Ewing stadi-altimeter: method of booking a traverse
46   Ewing stadi-altimeter: sketch of mounting on theodolite
47   Plane table traverse: correct plotting of detail
48   Plane table traverse: incorrect plotting of detail
49   Beaman arc: calculation of observations; comparison of scales
50   Beaman arc: calculation of observations
51   Plane table resection: Bessel's method
52   Plane table resection: solution of triangle of error
53   Compass traverse: method of booking
54   Compass traverse: adjustment of misclosure
55   Compass traverse: correct and incorrect adjustment of misclosure
56   Compass traverse: calculation of altitudes
57   Compass traverse: method of booking slope readings
58   Compass traverse: calculation for plotting of formlines
59   Compass traverse: plotting of formlines
60a The Brunton instrument used as a compass
60b The Brunton instrument used as an inclinometer
61a Aneroid barometer: continuous record and graph from 6-minute readings
61b Aneroid barometer: traverse on terrace fragments
62   Aneroid barometer: method of booking and calculation of control wave

63 Aneroid barometer: plot of approximate control wave: calculation of adjustment to booked readings: resultant errors
64 Quickset Level: component parts
65 Quickset Level: telescope graticule
66 Levelling Staves: patterns of graduation
67 Levelling Staves: telescope images
68 Quickset Level: effect of collimation error in altitude
69 Quickset Level: adjustment of collimation error
70 Levelling: method of booking and calculation
71 Levelling: profile for observations of fig. 70.
72 Levelling: method of booking by 'rise and fall'
73 Theodolite: controls
74 Theodolite: components of horizontal system
75 Theodolite: use of levelling screws
76 Theodolite: Lower Plate tangent screws
77 Theodolite: Upper Plate tangent screws
78 Theodolite: components of vertical system
79 Theodolite: alignment of split bubble
80 Theodolite: graduation of vertical circle
81 Theodolite: micrometer reading
82 Theodolite: internal micrometer readings
83 Theodolite: adjustment for collimation in altitude
84 Theodolite: error of collimation in azimuth
85 Sextant: component parts
86 Sextant: measurement of vertical angles
87 Sextant: geometry of observation with sea horizon
88 Sextant: measurement of horizontal angles
89 Sextant: parallax angle
90 Sextant: relationship between true and observed horizontal angles
91 Sextant: true horizontal angles corresponding to an observed angle of 60°
92 Abney clinometer: component parts
93 Abney clinometer: view through eyepiece
94 Abney clinometer: double vernier
95 Abney clinometer: geometry of observation
96 Aneroid barometer
97 Plane table and accessories
98 Mounting plane table
99 Plane table: adjustment of the telescopic alidade
100 Plane table: the Indian clinometer
101 Prismatic Compass: component parts
102 Prismatic Compass: the use of mirror on steep slopes
103 Declination and Right Ascension
104 Declination and Hour Angle

105   Altitude and Azimuth
106   The Astronomical Triangle
107   The Equation of Time
108   Latitude and solar altitude: solar declination zero
109   Latitude and solar altitude: solar declination +ve or −ve
110   Observed altitude in relationship to substellar point
111   Position line circle
112   Relationship of observed and calculated altitudes
113   Shift of position line for difference of altitude $h - h_c$
114   Location of observer's position in relation to Position Lines
115   Azimuth lines for sun observations
116   Plotting of solar azimuth lines and location of observer's position
117   Trigonometrical Resection: observer inside triangle of known stations
118   Trigonometrical Resection: observer outside triangle of known stations
119   Plane Table Resection: geometry of Bessel's method
120   Plane Table Resection: geometry of Bessel's method
121   Simple plane trigonometry: the basic triangle
122   Simple plane trigonometry: Eastings and Northings
123   Simple plane trigonometry: sines and bearings
124   Simple plane trigonometry: cosecant examples

Introduction

This is not intended to be a book for professional surveyors. It is a handbook principally for geographers, but also for geologists, botanists and others who require minimal competence in mapping while achieving adequate standards of accuracy for their varied purposes. It therefore sets out techniques which should, with a little practice, result in third-order surveys giving accuracy in mapping on scales to 1:1000, which is about the largest scale on which most field scientists will wish to work, and which can be obtained with the minimum of labour assistance. In general it is assumed that there will be not more than two unskilled assistants. It should be clear that with these limiting assumptions it is not possible to envisage the degree of accuracy normal for a professional surveyor or engineer, with a team of trained assistants.

The techniques included here should suffice to satisfy the needs of most field workers who may wish to add detail to a pre-existing and published map or who may need to map a small area in full. Such maps may be needed for a study of vegetation distributions or changes; for the pattern of small-scale geomorphological features such as shingle or sand patterns; for the detailed delineation of closely spaced contours in a study of slopes; for the layout of experimental plots in microclimatological studies; for detailed siting of historic material; for the distribution and heights of river terrace fragments; for the plan of an African village; for the pattern of prehistoric settlement; for the movement of glacier ice; and so on. It may be possible to map the detail required from air photographs if these are available, but an accurate map cannot be made from photographs without some ground control points giving both location and altitude. This book covers field techniques for complete surveys or for provision of adequate ground control for later laboratory mapping from air photographs.

Every survey needs to be executed to pre-determined standards of accuracy. An engineer or a professional surveyor often needs to determine geographical co-ordinates to the nearest 25 mm or better, because legal issues of land ownership are involved and co-ordinates must be precisely defined. The scientist in most cases requires a map as the end product of the survey, and the map scale determines the accuracy of his work. The *plottable error* is the degree of accuracy to which a measured length can be shown on the map. It is possible to draw accurately to the nearest 0.25 mm, so that to avoid map errors all ground measurements must be taken with sufficient refinement to meet this requirement. If the map scale is 1 : 1000, 1 mm on the map represents 1000 mm on the ground, and 0.25 mm on the map represents 250 mm on the ground, i.e. the plottable error is 250 mm and all measurements *on this scale* must be made to the nearest 250 mm. In practice it is usual to be slightly more accurate, i.e. measurements to the nearest 200 mm should suffice. If, however, the scale of the final map were 1 : 25,000, the survey would be executed in very different terms: 0.25 mm on the map would represent 6250 mm on the ground, and measurement to the nearest 5 m would be adequate. To measure to the nearest metre or tenth of a metre would be a waste of time, as on this scale a ground length of, for example, 263 m will be 10.5 mm on the map; a length of 262 m will also be 10.5 mm on the map; a length of 263.7 m, measured carefully to the nearest tenth of a metre, is still 10.5 mm on the map, and effort expended in refinement of measurement is therefore wasted. Indeed, on a scale of 1 : 25,000, minor measurements such as offsets can be paced with an accuracy within the plottable error provided that the observer's length of pace is known. The scale is therefore of fundamental importance in determining the method to be used.

Other factors which will affect the field worker's choice of method are

   (i)   ease of operation with a minimal labour force under different
         conditions
  (ii)   relative accuracy of different methods
 (iii)   cost or availability of equipment and the degree of expert knowledge
         required to operate it.

In most cases the scientist will not own the equipment which he uses. It is likely to be borrowed or hired, and it is difficult to obtain very refined instruments. Without adequate training and experience their use would in any case be unprofitable as well as unwise, and so advanced techniques are largely excluded here. A tellurometer or a geodimeter, for example, can give very precise measurements of length, but the average field worker will find it easier and quicker to observe angles over a small triangulation scheme than to hire experts to measure electronically the sides of the triangles. Electronic distance measurement (EDM) equipment is easy to use but very expensive, and it is most unlikely that the non-specialist will be able to borrow it; he may be able to hire it if he takes one or two days of instruction and practice, but he is likely to find the rate of hire prohibitive for any length of time if he is working with minimal resources. Also, the scale of map required may not justify expenditure on very highly accurate methods. With a small labour force it is generally easier to use optical measuring equipment than it is to measure linear distances accurately beyond a certain length. If, on the scale used, measurements to the nearest 0.25 m are within the plottable error (i.e. virtually all scales smaller than 1 : 1000) it is allowable to use optical measurements of distance for the fixing of detail. Nevertheless, the setting-out of initial fixed reference points should be done to a greater accuracy because the methods described allow work close to the limits of plottable error in the final stage of the work, i.e. the mapping of detail, and therefore assume no error in the reference framework.

General principles

Certain basic principles can be listed for any survey operation:

(1) Select in advance the scale to be used as this determines the plottable error and the degree of accuracy to be maintained in the taking of measurements.

(2) Always work from the more accurate to the less accurate methods, i.e. a compass traverse can be run between points fixed on a theodolite traverse, and adjusted accordingly—the converse is inadmissable. 'From the whole to the part' is an invaluable precept: each survey requires a reliable framework to which detail can be added, like bricks to a steel structure. One cannot create a safe building merely by adding brick to brick without previously established guide lines. Each survey should therefore be based on fixed points. These, or one of these, may be part of the national system, but usually they will be referred to an arbitrary origin and will be established for the particular survey in hand. It is highly desirable to fix each survey to the national system if possible.

(3) Orient each survey, preferably on Grid North if tied to the national system, otherwise on True North. If neither of these is possible, use Magnetic North. If no determination of any kind is possible at the start of a survey, work on an Assumed North and correct the map as soon as a more accurate orientation is obtained. It is insufficient to fix a survey to a single national point unless the survey is also oriented correctly. Determination of azimuth from observations to the sun is included in this book because these observations are relatively easy to make and to calculate, and the accuracy achieved is adequate for most geographical purposes.

(4) Plan all operations in terms of minimal labour. It is usually easier to acquire additional personal skill in use of instruments than it is to acquire additional helpers at little cost. This may involve slight departure from the standard professional procedures.

(5) Before starting work, insert an adequate number of fixed marks which will be included in the survey. Triangulation points, for example, should be semi-permanent in character (e.g. a length of piping, or a very large nail, hammered into bedrock; a 2-metre length of piping hammered into marsh until only 20 mm is exposed above the surface) so that they can be used in subsequent years for check surveys if necessary. Theodolite traverse points should all be marked by wooden pegs hammered into the ground prior to observation, so that they can be used in later stages of the survey. It is not necessary to insert such definite marks at stations on a compass survey, which is unlikely to form the basis for later work, but temporary pegs are useful in contouring and plane-tabling.

(6) Every part of a survey must include checks. It is essential for a surveyor to acquire an attitude of mind such that he never proceeds without a check on the accuracy of his previous work. This shows itself in such ways as the automatic checking of a plane-table resection by reference to a fourth point if available, or the constant counting of paces while walking between traverse points, as an approximate check on recorded length measurements. It involves closed circuits in levelling, or continuous reading of stadia lines as a check on the crosswire reading, even though the stadia readings may not be recorded but are used only for mental arithmetic on the spot. It involves the use of forms of computation which check against errors in the calculations. This attitude of mind has, in the past, been only too rare with geographers and others, who have frequently displayed unjustifiable confidence in unchecked data based on faulty arithmetic. It is to be hoped that the modern tendency towards use of quantitative methods will encourage closer attention to checks on accuracy. Surveying is probably the oldest form of quantitative observation which geographers have been making. If surveying techniques are employed correctly, the resulting map or profile is a statistical model which cannot be faulted within the limitations of scale selected: only if the map is enlarged does it become inaccurate as the degree of refinement of measurement employed in the survey is then inadequate for the plottable error of the new scale.

The structure of the book

As this book is intended to be a guide or instruction manual for the non-specialist, it is set out in the form considered to give the logical order of working for anyone making a complete survey of an area of perhaps five square kilometres, with no adequate base map already available. In such a case it is necessary to start with a positional fixation, but it is assumed that most scientists will be working in an area which has at least one co-ordinated mark available, and so simple astronomical fixation has been included only as an appendix. The book therefore starts with orientation observations at a known point (as only one fixed point may be readily accessible) and continues with the lay-out, observation and calculation of control networks to provide fixed points, at requisite frequency, on which detailed mapping can be based.

If the area to be mapped is too small to justify a triangulation network, an adequate density of control points may be inserted by traverses, and anyone who wishes may start at that point. If only a small area is being mapped, it may be possible to survey all necessary detail from only one or two instrumental positions, and a start can be made at the section on plane table work, ignoring earlier chapters. Essentially, therefore, the more difficult and more accurate methods of working come at the beginning of the book, just as in the field they come at the start of a survey.

A description of instruments likely to be used is given separately from the instructions for observatonal procedures, and the worker unfamiliar with the equipment should study the appropriate accounts and should acquire some competence in handling the instruments before attempting to use them on a project.

The appendices include some very simple trigonometry for those students who have not taken mathematics beyond basic standards, and an account of 7-figure logarithmic tables of trigonometrical functions. The advent of small battery-powered pocket calculators which incorporate trigonometrical functions has made survey calculations very much easier and quicker, but it is highly desirable that anyone whose recent mathematical experience has been limited should calculate the first part of a survey with logarithmic tables in order to regain familiarity with trigonometrical functions and so avoid errors of misinterpretation of calculator readings.

The tables at the end of the book can be used in the field or in the laboratory calculations, and should be adequate within the limits of accuracy likely to be sought by the field scientist.

Typical techniques to be used

There are no hard and fast rules about techniques to be used. Any survey can usually be done accurately by several different methods and the surveyor will choose his method in the light of equipment available and his own experience. However, the following summary may assist the inexperienced in deciding which method to adopt.

In general, a network of fixed points (chapter 1) will be required only for areas greater than half a kilometre in diameter. A traverse to establish some fixed points (chapter 2) will usually be necessary when the area exceeds 200 m in diameter, unless all detail to be mapped can be seen from one point. If the ground is very uneven or the vegetation dense, a network or some traverse control may be required for smaller surveys also.

For small detail surveys, chapter 3 may be sufficient (albeit using the tables at the end of the book and referring to relevant sections of chapter 4). Chapter 5 is intended for approximate location of points on expeditionary work, but more particularly for orientation of surveys by sun azimuths.

Scale	Map shows accuracy on ground to nearest:	Size of area	Examples of purpose of map	Method	Number of workers	See pages
Areal mapping						
1: 50 or 1:100	12.5 mm 25 mm	Radius < 100 mm	Successive plots of melting snow patches sand accumulation seasonal vegetation growth etc.	Lay out triangulation points at intervals not exceeding 50 metres, measured to nearest 5 mm.	2 minimum	14
				Detail by offsets from lines between points. Plot in drawing office.		54
		Radius >100 m		Triangulation to be observed with a 20″ theodolite (or better).	2 minimum	14
				Detail and plotting as before.		54

Note. Truly accurate measurement is difficult on these very large scales.

Scale	Map shows accuracy on ground to nearest:	Size of area	Examples of purpose of map	Method	Number of workers	See pages
1: 500 or 1: 1000	125 mm 250 mm	Radius <100 m	*Detail mapping* Vegetation distribution Location of boulders on a talus slope. Rill patterns Beach cusps etc.	Plane table. Detail by rays and distances.	2 minimum	80
		Radius >100 m		Plane table. Detail by rays and distances from positions established by		80
				triangulation	2 minimum	14
				theodolite traverse	3	47

			Height information
			Contouring
			  Vertical interval
			  2 m or less
				Plane table and quickset level	2 or 3	94

			Formlines	Interpolation between spot heights fixed tacheometrically with telescopic alidade or theodolite	2 or 3	85 96

Scale	Map shows accuracy on ground to nearest:	Size of area	Examples of purpose of map	Method	Number of workers	See pages
1: 2000 or smaller	500 mm or more	Radius <200 m	*Detail mapping* Vegetation distribution Landform surveys etc.	If whole area visible from a central point: Plane table and telescopic alidade	2	85
				Plane table and theodolite tacheometry	2 or 3	96
				If not, as for Radius >200 m		
		Radius >200 m		As above, but from successive positions previously established by		
				triangulation	2 minimum	14
				theodolite traverse	3	47
				plane table traverse (for scales of 1: 5000 or smaller)	3 minimum	80
				As above, but from positions fixed by plane table resection from triangulation or other points already established	2	89
			Height information Contouring	As for 1:1000	2	
			Formlines	As for 1:1000, or by plane table height resection using Indian clinometer	3 2	94

Scale	Map shows accuracy on ground to nearest:	Size of area	Examples of purpose of map	Method	Number of workers	See pages
1:10000 or smaller	2.5 m		Detail mapping and Height information	Successive plane table resection from points previously established by	1 or 2	89
				triangulation	2	14
				theodolite traverse	3	47
				Detail around resected points by pacing provided that error on distance does not exceed 2.5 metres		
				Formlines by interpolation between resected spot heights with Indian clinometer		94
Linear mapping						
1:50 or 1:100	12.5 mm 25 mm		Profile measurement	Quickset level along taped lines	2	43/125
1:500 or 1:1000	125 mm 250 mm		Profile measurement Detail alignment rills intrusive dykes etc.	As above Theodolite traverse with offsets	2 3	47
1:2000 or smaller	500 mm or more		Stream channels etc.	Theodolite traverse with offsets Plane table traverse	3 3	47 80
1:10000 or smaller	2.5 m or more		Low water mark Geological boundaries Stream channels Formlines etc.	Plane table traverse Plane table resection from points previously established by	3 1	80 89
				triangulation	2 minimum	14
				theodolite traverse	3	47
				Prismatic compass traverse between established points	2 or 3	97
			Spot heights	Aneroid traverse	1 or 2	111
Locational fixation						
			Orientation	Sun azimuth	1	2
			Latitude	Sun altitudes	1	2/177
			Latitude and longitude	Position lines	1	179
Offshore observations						
			Location of Observer's boat	Collins Point resection by sextant	1 or 2	189
			Angular measurement from boat	Sextant (can also be used for measurement of angles where all points observed are at approximately the same level)	1 or 2	76

1 Provision of fixed marks

If the area to be surveyed is larger than half a kilometre in diameter, it cannot be mapped from a single point. In such a case it is necessary to set out a framework of fixed points, whose relative positions have been determined with accuracy exceeding that of the plottable error of the intended map. The excess accuracy allows for a map of larger scale to be based upon the same framework if this proves desirable at a later stage. For example, a map of an area marginal to a glacier might be made on a scale of 1:5000, but a detailed plan of outwash patterns, on a scale of 1:1000, might be required for a part of the area. Another example would be a map of broad vegetation zones on a scale of 1:2000, but with marginal areas surveyed in great detail on 1:500. Fixed point frameworks are therefore usually surveyed with care. In some cases, such as a small plane-table survey, it may be sufficient to measure a short base-line only, and to develop detail from this, but in many surveys it is essential, or at least desirable, to use a theodolite to establish a number of marks with relatively high accuracy. The marks may form part of a minor triangulation scheme or may be stations in a theodolite traverse.

Orientation

It is essential that the survey should be oriented. If it is possible to establish an instrument over a fixed point in the national survey and to sight to a similar point, then the new work can be tied in without difficulty, provided that the bearing between the two fixed points is known or can be calculated from their co-ordinates. However, in most cases only one fixed point will be available, and this will not necessarily be a mark over which an instrument can be centred. It may, for example, be a bench mark on a wall. Provided that its national co-ordinates are recorded, a survey can be tied to it if the survey can itself be oriented.

Sun azimuths

The simplest method of accurate orientation available for tying a survey to a single known point is the observation of the sun and the calculation of azimuth. This implies reasonable familiarity with the theodolite controls (see pp. 129–30). The method used is known as observation of azimuth by ex-meridian altitudes of the sun. Essentially it consists of observing the altitude of the sun at known times (which can be checked in the calculation), together with the simultaneous measurement of the horizontal angle at the observer between the telescope pointings to the sun and to some Reference Object. It is then possible to solve the astronomical triangle (see p. 174) to obtain the azimuth of the sun at the time of each observation (fig. 1). The *azimuth* of the sun is the true bearing of the sun, i.e. the angle at the

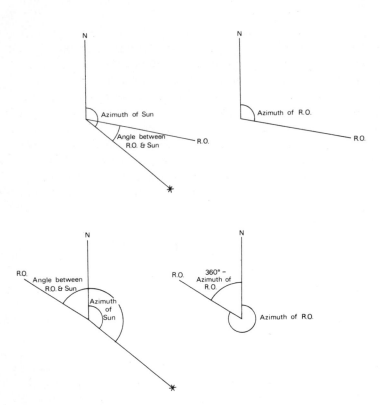

Fig. 1   Relative positions of sun and Reference Object in the calculation of azimuth.

observer between the direction of True North and the direction of the sun. The azimuth of the Reference Object is its true bearing, i.e. the angle at the observer between the direction of True North and the direction of the R.O.

The astronomical triangle can be solved for the azimuth angle of the sun by the use of the formula

$$\tan \frac{A}{2} = \sqrt{\sec s . \sin (s - b) . \sin (s - \phi) . \sec (s - p)}$$

where

$$s = \frac{b + \phi + p}{2}$$

b = the altitude of the sun

ϕ = the latitude of the observer

p = the polar distance of the sun,

i.e. the angular distance from the sun to the elevated pole. When the sun is overhead at the equator (i.e. when the sun's declination is zero) the north polar distance is clearly 90°. When the sun is north of the equator it is said to have a north declination δ, and the north polar distance is 90° $-\delta$; when the sun has a south declination δ the north polar distance is 90° $+\delta$. The sun's declination is tabulated for the year, and can be calculated for the time of each observation.

The Nautical Almanac, published annually by H.M.S.O., gives the declination of the sun for every hour of each day. For example, for October 16th, 1973, it lists

G.M.T.		Dec.	
d	h	°	′
16	00	S. 8	45.0
	01		46.0
	..		..
	..		..
	..		..
	14		57.9
	15		58.9
	16	8	59.8
	17	9	00.7
	18	9	01.6
	..		..
	..		..
	23		06.2

If the sun were observed at $16^h \ 42^m$ G.M.T., the sun's declination at that time can be interpolated.

At $16^h \ 00^m$	$\delta = $ S. $08° \ 59' \ 48''$
$17^h \ 00^m$	$\delta = $ S. $09° \ 00' \ 42''$
Change in δ for 60 minutes	$= $ S. $\qquad 54''$
Change in δ for 42 minutes	$= $ S. $\dfrac{42}{60} \times 54''$
	$= $ S. $\qquad 38''$
At $16^h \ 00^m$	$\delta = $ S. $08° \ 59' \ 48''$
Increment for 42^m	$= $ S. $\qquad 38''$
At $16^h \ 42^m$	$\delta = $ S. $09° \ 00' \ 26''$

Note that the final figure, although given to seconds, has an accuracy only of $\pm 6''$, this being the accuracy of the tabulated data.

An alternative source of data is the H.M.S.O. publication "Star Almanac for Land Surveyors". This lists the sun's declination for every 6 hours, and a similar interpolation can be calculated.

For any one observation the altitude of the sun (h) is measured with the theodolite; the latitude of the observer (ϕ) is known approximately from any base map; knowing the time of the observation it is possible to calculate the sun's declination and therefore the polar distance (p). From these three variables it is possible to find s, and hence to calculate the azimuth angle of the sun.

It can be seen from fig. 1 that if the azimuth of the sun is calculated, and the angle between the R.O. and the sun is measured, the azimuth of the R.O. can also be computed. It is always desirable to draw a little sketch similar to fig. 1, to show the relative positions of the R.O., the sun and the north point, as this assists the latter stages of the calculation.

The non-mathematician need not be alarmed by the formula and need not concern himself with its derivation. The calculation requires little more than the ability to add and subtract and the ability to use tables of logarithms. He should, however, note the precaution stated on p. 13 that observations for azimuth should not be taken when the sun is near the observer's meridian, as large errors are then liable to enter the calculations and to make them insoluble.

Observation   The theodolite is set on its tripod and is centred over a station mark, which may be the end of a vertical length of metal pipe, the head of a nail, or a circular hole made in the top of a wooden peg with the tip of a chaining arrow. If centring has been done with a plumb bob, the theodolite is then clamped on its base and is levelled with the footscrews; if an optical centring device is used, the instrument is levelled first and is clamped after centring is completed. The lower plate (the graduated horizontal circle) is then clamped also. The eyepiece is adjusted to the observer's vision to show clear crosswires, and the telescope in Face Left position is pointed at the Reference Object and clamped. The R.O. should be a clearly defined mark which can be readily and repeatedly intersected without difficulty. It may be a chaining arrow held against a Bench Mark; it may be a survey pole held on a traverse or triangulation station; it may be the top of a church spire, or any other suitably rigid point. It should be sufficiently remote from the observer for intersection by the vertical cross-wire to be clear and unambiguous. If necessity compels the use of the next traverse station, and a survey pole held thereon appears in the telescope with considerable width, so that its centre has to be estimated, substitute an arrow for the pole so that there is no doubt that exactly the same point is being intersected at each observation to the R.O. Final intersection of the R.O. is effected with the upper plate tangent screw in the usual manner. If the R.O. is reasonably remote, adjustment of focus when changing from R.O. to sun, or vice versa, will be minimal.

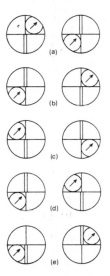

Fig. 2   Intersections of the sun's image on opposite Face positions of the theodolite telescope. Conditions (a) and (b) are identical; (a), (b) and (c) would give an acceptable mean reading; (e) is less desirable; (d) would give an erroneous result. Half the vertical line is split in many theodolites, as it is frequently easier to judge when a thin target mark is midway between two lines than it is to be sure that it is exactly bisected by a single line.

The horizontal circle reading for the R.O. is booked, and the dark glass supplied with the theodolite is fitted to the eyepiece. The upper plate is released, the telescope is pointed at the sun and the upper plate and the vertical circle are clamped. It is not possible to intersect the apparent centre of the sun, as the sun's image is too large, so observations need to be made to opposite limbs of the sun in quick succession, and the mean value taken. To do this the observer has to achieve a pair of observations as in fig. 2(a) and 2(b). The sun moves rapidly past the crosswires, and the observer keeps the vertical crosswire against the edge of the sun's image with the aid of the horizontal (upper plate) tangent screw, which he turns slowly, allowing the image to move vertically until the desired limb is exactly on the horizontal crosswire. The observer then records, in order, the time, the displacement of the ends of the bubble in the precise spirit level, the vertical circle reading and the horizontal circle reading. He then reverses the telescope, and on

Face Right observes the sun again, using the opposite quadrant as in fig. 2(a) or 2(b), and books the new values. The dark glass can then be removed and the observer intersects the R.O. and records the horizontal circle reading.

It is important that only conditions (a), (b) or (c) in fig. 2 should be observed. Condition (d) will obviously give erroneous results, as the mean position of the two images would not be intersected by the crosswires at the mean time, but would lie to one side of the vertical axis. Condition (e) would be less disastrous, but is undesirable. The instrument illustrated has the vertical axis divided as described on p. 5, and with this type of graticule the quadrant used must be bounded by single lines.

The maximum allowable time interval between the Face Left and Face Right observations to the sun is 4 minutes. This is because the sun's trajectory is a curved and not a straight line (fig. 3). The mean of the two

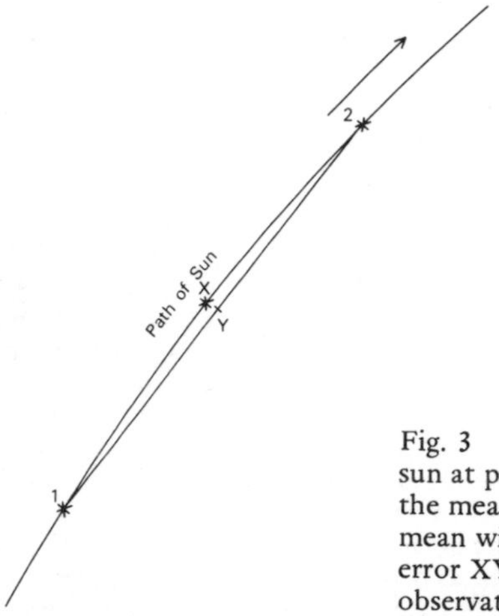

Fig. 3   Observations on different Face are taken to the sun at positions 1 and 2. The desired true position for the mean of the two times is at X, but the arithmetic mean will be at Y, giving an erroneous value. To keep the error XY acceptably small, the time interval between observations 1 and 2 must not exceed 4 minutes.

observations therefore lies at a false position *Y* on a chord, and not at the true position *X* on the arc of the trajectory, and if the time interval is too great the divergence between arc and chord introduces errors of unacceptable size. Because the time interval between F.L. and F.R. observations is restricted, rapid intersection of the sun for the second sighting is important. It is not always easy to achieve this. Obviously one cannot line up the open telescope sights with the sun as would be done with observations to opaque objects. Many observers find that the most rapid method of alignment is by moving the telescope to obtain coincidence of the shadows of the open sights on a hand held near the eyepiece, and clamping in this position. The sun should then be in the field of view.

Accurate observation through the telescope becomes difficult if the altitude exceeds 45°, and a right-angled eyepiece has to be introduced above this value. For most scientists without specialist training this involves an additional and unfamiliar hazard likely to impair the quality of the observations, and in many cases the extra eyepiece will not be available. It is therefore advisable to take sun observations early in the morning and late in the afternoon at seasons when the sun is too high in altitude for a large part of the day. If possible, take both morning and afternoon observations when the sun is at approximately the same altitude, in order to balance and eliminate unknown errors introduced by unusual conditions of atmospheric refraction.

For an azimuth to be acceptable for minor surveys of this nature, it is sufficient to have three calculated values agree within 30 seconds of arc. Weather permitting, it is customary to take 5 complete sets of observations, each recording R.O./ ☉ / ☉ /R.O. and to calculate these simultaneously. Three of the five can be expected to agree within the prescribed limit.

Calculation　A useful computation form is given as fig. 5, using observations booked as in fig. 4.

The horizontal angle from the R.O. to the sun is calculated in the observation book, and the mean taken of Face Left and Face Right angles: unlike most theodolite angles, these will not agree to within seconds, on account of the horizontal movement of the sun between the two observations. Each vertical angle is then worked out, corrected if necessary for dislevelment of the bubble. An older instrument normally has a spirit level with a stated value

Observer J. C. Pugh　　Date 11.3.73　　Survey Ambersham Common

AT	TO	FACE	HORIZONTAL CIRCLE ° ′ ″	HORIZONTAL ANGLE ° ′ ″	VERTICAL CIRCLE ° ′ ″	VERTICAL ANGLE ′ ″	TIME	LEVEL E O	REMARKS
△H	G	L	147 48 15						
	☉		81 20 50	293 32 35	111 27 22	21 27 22	15 30 50	—	
	☉	R	261 15 25	293 27 05	69 20 39	20 39 21	15 33 00	—	
	G		327 48 20	293 29 50		21 03 21	15 32		
	G	L	181 11 26						
	☉		115 41 15	294 29 49	110 35 44	20 35 44	15 37 40	—	
	☉	R	296 50 01	295 38 41	70 15 38	19 44 22	15 40 05	—	
	G		01 11 20	295 04 15		20 10 03	15 39		
	G	L	213 50 08						
	☉		150 29 49	296 39 41	109 40 27	19 40 27	15 44 40	---	
	☉	R	330 25 56	296 35 23	71 09 16	18 50 44	15 47 00	—	
	G		33 50 33	296 37 32		19 15 35	15 46		
	G	L	256 20 35						
	☉		194 03 10	297 42 35	108 42 01	18 42 01	15 52 00	—	
	☉	R	15 05 00	298 44 12	72 06 25	17 53 35	15 54 00	—	
	G		76 20 48	298 13 23		18 17 48	15 53		
	G	L	292 23 20						
	☉		232 13 05	299 49 45	107 44 52	17 44 52	15 59 00	—	
	☉	R	52 09 59	299 46 36	73 06 16	16 53 44	16 01 20	—	
	G		112 23 23	299 48 11		17 19 18	16 00		

Fig. 4　Booked observations for 5 sets of sun azimuths. The theodolite used had a split bubble image so that no E and O readings of the spirit level were required.

Survey *Ambersham Common*   TIME & AZIMUTH

Date *11. 3. 73*

Latitude *50° 58' 00" N.*   Station *H*

Longitude *00° 41' 50" W.*   R.O. *G*

Observer *J. C. Pugh*   Magnetic *307½°*
                                Bearing

Insert directions to Sun and R.O.

Watch Time		h m s 15 46	Watch Time G.M.T.	h m s 15 31	h m s 15 38	h m s 15 45	h m s 15 52	h m s 15 59
Mean Refraction	−	°02' 46"		°02' 31"	°02' 38"	°02' 46"	°02' 55"	°03' 05"
Parallax	+	08		08	08	08	08	08
Total Correction	−	02 38		02 23	02 30	02 38	02 47	02 57
Mean Obs.Altitude		19 15 35		21 03 21	20 10 03	19 15 35	18 17 48	17 19 18
True Mean Alt. (h)		19 12 57		21 00 58	20 07 33	19 12 57	18 15 01	17 16 21
Declination of Sun		S.03 35 32		03 35 46	03 35 39	03 35 33	03 35 26	03 35 19
Polar Distance (p)		93 35 32		93 35 46	93 35 39	93 35 33	93 35 26	93 35 19
Altitude	h	19 12 57	h	21 00 58	20 07 33	19 12 57	18 15 01	17 16 21
Latitude	φ	50 58 00	φ	50 58 00	50 58 00	50 58 00	50 58 00	50 58 00
Polar Distance	p	93 35 32	p	93 35 46	93 35 39	93 35 33	93 35 26	93 35 19
Sum		163 46 29	Sum	165 34 44	164 41 12	163 46 30	162 48 27	161 49 40
÷ 2 = s		81 53 15	÷ 2 = s	82 47 22	82 20 36	81 53 15	81 24 13	80 54 50
	s − h	62 40 18	s − h	61 46 24	62 13 03	62 40 18	63 09 12	63 38 29
	s − φ	30 55 15	s − φ	31 49 22	31 22 36	30 55 15	30 26 13	29 56 50
	s − p	11 42 17	s − p	10 48 24	11 15 03	11 42 18	12 11 13	12 40 29
log cos	s	9.149 5800	log sec s	0.901 3009	0.875 3763	0.850 4200	0.825 4372	0.801 5664
log sin	(s−h)	9.948 6039	log sin (s − h)	9.945 0170	9.946 8076	9.948 6038	9.950 4711	9.952 3239
log cosec	(s−φ)	0.289 1609	log sin (s − φ)	9.722 0525	9.716 5559	9.710 8391	9.704 6564	9.698 2764
log sec	(s−p)	0.009 1259	log sec (s − p)	0.007 7711	0.008 4273	0.009 1263	0.009 8992	0.010 7142
Sum		19.396 4706	Sum	0.576 1415	0.547 1671	0.518 9892	0.490 4639	0.462 8809
÷ 2 − log tan t /2		9.698 2353	÷ 2 = log tan A/2	0.288 0708	0.273 5836	0.259 4946	0.245 2320	0.231 4405
t/2		26° 31' 35"	A/2	62° 44' 42"	61° 57' 34"	61° 10' 53"	60° 22' 47"	59° 35' 30"
t (arc)		53 03 10	A from N (am = clockwise)	125 29 24	123 55 08	122 21 46	120 45 34	119 11 00
÷ 15 = t (time)		h m s 03 32 13	(pm = anticlock- wise)					
Before noon subtract from 24	h		Azimuth of Sun (clockwise from N)	234 30 36	236 04 52	237 38 14	239 14 26	240 49 00
− L.A.T								
Equation of Time ±		+ 10 04						
− L.M.T.		03 42 17	Horiz Angle R.O./Sun	293 29 50	295 04 15	296 37 32	298 13 23	299 48 11
Longitude (time) W +		+ 02 48						
E − − G.M.T.		03 45 05	Azimuth of R.O.	301 00 46	301 00 37	301 00 42	301 01 03	301 00 49
Watch Time		03 46 00						
Watch Error		+ 00 55						

Equation of Time is ZERO on	Apr 15 Jun 15 Aug 31 Dec 24
and is	+ zero − zero + zero − zero +

```
                    301  00  46
                         00  37
                         00  42
                         01  03
                         00  49
                    ───────────
                    301° 00' 47"
```

Fig. 5 Calculation of the five azimuth sets observed in fig. 4. The first column calculates the G.M.T. of the middle set (the third of the five), and shows the watch time to be 55 seconds fast. All five azimuths are then calculated on corrected watch times. As all sets agree within 26", the mean of the five is taken as the correct azimuth of the Reference Object (Station G) from the observer's position at Station H.

in seconds for dislevelment by one graduation, and the vertical angle can be amended accordingly. For example, fig. 6(a) shows dislevelment of a 5 second bubble. If the longer graduation is called 10, with numbers increasing outwards from the centre of the tube, the object glass end reads 8 and the eyepiece end reads 11. Difference = 3. Dividing by 2 for the two ends of the bubble, it can be seen that dislevelment equals 1.5 graduations of 5 seconds each = 7.5 seconds. (Dislevelment is 1.5 and not 3. If the bubble is moved by 1.5 graduations from E towards 0, both ends will read 9.5 and the bubble

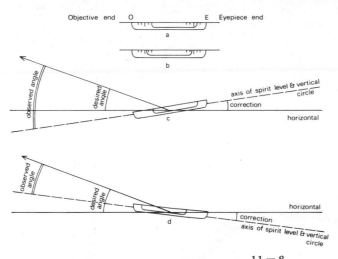

Fig. 6 (a) E reads 11: 0 reads 8. Dislevelment = $\dfrac{11 - 8}{2}$ = 1.5.

(b) Lateral displacement of 1.5 divisions gives a central bubble.
(c) Bubble is displaced towards the observer. Subtract the correction from the observed angle.
(d) Bubble is displaced away from the observer. Add the correction to the observed angle.
(Dislevelment angle is greatly exaggerated.)

will be central, as in fig. 6(b)). As the bubble is at the eyepiece end, the horizontal line of collimation is tilted downwards away from the observer, thus increasing the observed vertical angle. The 7.5 seconds correction has therefore to be subtracted from the observed vertical angle. If the 0 reading exceeds the E reading the correction is added to the observed angle. Modern theodolites tend to be fitted with a reflector showing a split bubble (see p. 138) and this is brought to centre by the observer immediately before taking each sun observation, so that there is no need for adjustment of the vertical angle. The mean of the corrected vertical angles is also entered in the observation book.

The first step in the calculation is the working out of the first column of the form in fig. 5. The observer enters data for one of the middle sets in his series of observations, and calculates not the azimuth angle but the hour angle of the sun, to check his watch. This is an example of the checking of values which was mentioned on p. xviii. The astronomical triangle is solved for the hour angle t, where

$$\tan \frac{t}{2} = \sqrt{\cos s . \sin (s - b) . \operatorname{cosec} (s - \phi) . \sec (s - p)}$$

and

$$s = \frac{b + \phi + p}{2}$$

as before.

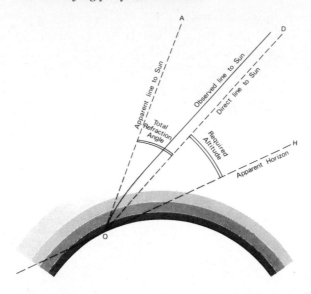

Fig. 7   The incoming light ray is refracted by increasing atmospheric density. The observer at *O* therefore measures altitude as the vertical angle *AOH* instead of the required angle *DOH*. *AOD* is the refraction angle, and must always be subtracted from the observed angle. (Refraction angle is greatly exaggerated.)

Working down the column, compensation for atmospheric refraction of the light rays is obtained from table 2. Clearly refraction will be zero if the telescope points vertically upwards, and will be greatest when the sun is at the horizon, as refraction results from the variation in the density of the

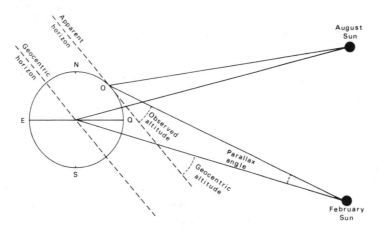

Fig. 8   The parallax angle + the observed altitude = the geocentric altitude. The parallax correction must therefore always be added to the observed altitude. This is true irrespective of seasonal variations in the sun's declination. (Parallax angle is greatly exaggerated.)

atmospheric layers through which light rays pass. The greater the sun's altitude the less is the refraction error but the greater is the difficulty of observation without a right-angled eyepiece. The best compromise is to observe the sun at altitudes of about 40°, when the refraction error is no more than about 1 minute of elevation and the observer can work without strain. It can be seen from fig. 7 that the effect of refraction is to make the altitude of a heavenly body appear greater than it should, and the correction for refraction is therefore always negative. In precise surveys a further adjustment of refraction is made to allow for variation with temperature and with barometric pressure, but these corrections amount to no more than a few seconds of arc, and can be ignored here as being smaller than the precision of the instruments likely to be used by the non-specialist scientist.

The second correction is for parallax. The Nautical Almanac, the Star Almanac and similar lists give information for the sun with reference to the centre of the earth and therefore to a true horizon through the centre of the earth. The observer uses the apparent horizon as illustrated in fig. 8, and there is a small difference between the observed altitude and the geocentric altitude, depending on the amount by which the observer is displaced from the direct line from the centre of the earth to the centre of the sun. This difference is known as the parallax correction, and must be added to the observed altitude to obtain the geocentric altitude. Parallax does not exceed 9 seconds, and for altitudes of about 40° is approximately 7 seconds. The correction is large enough to justify adjustments of mean angles observed with a theodolite graduated to 20 seconds of arc. Correction values for parallax are given in table 1.

With these two corrections entered in the form, the True Mean Altitude (h) is obtained. The declination of the sun is tabulated at 1-hour intervals in the Nautical Almanac, and the value for the date and time of observation is entered in the form. In making this calculation, it must be remembered that the Nautical Almanac uses Greenwich Mean Time, and watch times should be altered accordingly before extracting values from the tables. The Polar Distance (p) refers to the elevated pole, i.e. if the observer is in the Northern Hemisphere he will use the polar distance from the North pole. When the sun's declination is North (i.e. in the period March 21st to September 23rd) the North Polar Distance is obtained by subtraction of the north declination from 90°. For the rest of the year, when the sun's declination is South, the North Polar Distance is obtained by addition of the south declination to 90°. The latitude (ϕ) being known from base maps, the values of h, ϕ and p are entered on the form, and values calculated for s, $s - h$, $s - \phi$ and $s - p$. The logarithms of the trigonometrical functions are then taken from a book of tables such as Shortrede (see p. 196).

Log cos s and log sin $(s - h)$ are read directly from the tables. Log cosec $(s - \phi)$ is obtained by subtracting log sin $(s - \phi)$ from zero, and log sec $(s - p)$ by subtracting log cos $(s - p)$ from zero. Strictly speaking one subtracts from ten and not from zero, as all logarithms are kept positive, e.g.

To obtain log sec 11° 42′ 17″

log cos 11° 42′ 17″ =   9.990 8741

                         10.000 0000
                       − 9.990 8741
                         _____

log sec 11° 42′ 17″ =   0.009 1259

The four logarithms are added, and the sum divided by two to obtain the logarithm of the square root.[1] This product is the logarithm of the tangent of the angle which is half the hour angle t. A search is made in Shortrede for this logarithm, and the appropriate angle entered in the form and doubled. The Equation of Time (see p. 175) is also entered and Local Mean Time calculated. If the observer enters longitude expressed in time as a correction, the error of the watch on G.M.T. is then found. Longitude must be included as a correction in time calculations, as the sun when crosssing the Greenwich meridian at Greenwich Noon will not yet have reached the meridian of a point further west, where Local Noon will await the sun's crossing of the local meridian. For example, a point at longitude 01° 00′ W. will have a Local Time of 11.56 a.m. when it is Noon at Greenwich, the difference of 4 minutes being the time taken by the sun to travel through 1 degree of longitude. Strictly speaking, of course, it is the time taken by the earth to rotate through 1 degree of longitude. In 24 hours it rotates through 360°; in 1 hour through 15°; in 4 minutes through 1°, etc.

   The times of all observation sets are then corrected by any error disclosed. All remaining columns can then be calculated simultaneously for azimuth. Refraction and parallax corrections are applied to all sets and True Mean Altitude calculated for each. (This has, of course, already been done for the set used in the Time calculation.) Declination of the sun is worked out for each set (the set used in the calculation of Time may now need adjustment of this value to the corrected time of observation) and the Polar Distances are calculated. Values for h, ϕ, and p are entered, s is found, and the values for $s - h$, $s - \phi$ amd $s - p$ are inserted. Logarithms are then entered from the tables. Log sin $(s - h)$ and log sin $(s - \phi)$ are taken directly; log sec s and log sec $(s - p)$ are obtained by subtracting the corresponding log cosines from zero. The logarithms are summed and the total halved to obtain the square root. This is the logarithm of the tangent of an angle

---

[1] "Division by two" is not always simple, nor is addition of the 4 logarithms. A study of the calculations in fig. 5 shows in the Time column the addition of 4 logarithms which should total 19.396 4706, but this is written as 9.396 4706, following the customary practice of removing unnecessary multiples of ten from the characteristic of the logarithm. Half of 9.396 4706 is written not as 4.698 2353 (which would be the tangent of a very small angle close to 1″) but as 9.698 2353, because it is really 19.396 4706 which is being divided by two.
   Similarly in the first of the azimuth columns, addition of the 4 logarithms would give a total of 20.576 1415, and half of this would be 10.288 0708, which is the figure used when searching in the tables to identify the angle $A/2$.

which must be doubled to give the azimuth angle of the sun at the time
of observation. The diagram at the top of the form assists the observer in
deciding which is correct of the possible angles which have the log tangent
obtained (remembering that this angle is half the azimuth angle and not
the azimuth itself). The horizontal angle from the R.O. to the sun is
subtracted from the sun azimuth to obtain the azimuth of the R.O. If
necessary, 360° is added to the sun azimuth before the subtraction, in
order to keep positive the angle finally obtained, as seen in fig. 9. Of the
calculated answers, at least three should agree within 30 seconds, and the
mean of these can be accepted. It is customary to observe at least four,

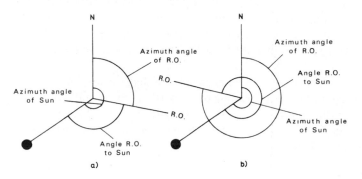

Fig. 9   (a) azimuth of R.O. = (azimuth of sun) − (angle R.O. to sun).
(b) azimuth of R.O. = (azimuth of sun + 360°) − (angle R.O. to sun).
The azimuth angle is always measured clockwise from North. The angle from R.O. to sun
is always measured clockwise also.

and preferably five sets in order to ensure sufficient agreement even
though accidental error may invalidate one set of observations. Five sets
can be observed in about 30 minutes.

   If the answers show lack of agreement, arithmetic is easily checked for
gross error by examination of neighbouring columns. On each line, values
or logarithms should change slowly. If there is no obvious arithmetical
error, and the azimuth values obtained show a fairly regular but overlarge
change in one direction, it is probable that the wrong angle has been taken
from the tables as the antilogarithm of log tan $A/2$.

Caution   The surveyor should note one important precaution. It was stated
on p. 7 that ideal altitudes will be as large as possible provided that they do
not exceed 45°. One exception to this is that observations for azimuth should
not be taken near mid-day, even if the altitude is low. For example, in
November or December in London, mid-day altitude of the sun may be
approximately 19°, the sun's declination being S. 19° and the latitude
roughly 51° 31′ N. In the calculation one then has

$$\tan \frac{A}{2} = \sqrt{\sec s \cdot \sin (s - h) \cdot \sin (s - \phi) \cdot \sec (s - p)}$$

with $A/2$ very close to $90°$ (as the azimuth angle is close to $180°$ from North) and $s = \frac{1}{2}(19° + 51° 30' + 109°) = \frac{1}{2}(179° 30') = 89° 45'$. The log tangent and log secant vary very rapidly indeed, so that a small difference in altitude may make a difference ten times greater in the calculated azimuth angle; any error in the latitude value used, or any unsuspected refraction condition can therefore greatly alter the value of the azimuth angle calculated, e.g. if one uses $51° 30'$ N. as the latitude value instead of a more accurate figure of $51° 31'$ N., the value of s used in the calculation will be too low by $30''$. A difference of $30''$ when s is in the vicinity of $89° 52'$, can make a difference in the logarithm of $0.030\ 0000$. If $A/2$ is $85° 50'$, a difference of $0.030\ 0000$ in the log tangent represents a difference of 17 minutes, and the azimuth varies by 34 minutes. Under such conditions it is preferable to take low angle observations away from the meridian, as errors due to greater refraction will be smaller than errors arising in the computation.

Care should therefore always be taken to ensure that s is not near $90°$, particularly if using a theodolite reading only to 10 or 20 seconds of arc. One should ensure that the variation per second of arc of any logarithm in the right-hand side of the equation is never more rapid than the variation per second of log tan $A/2$.

Other methods of orientation

The field scientist may hesitate to attempt a sun azimuth, or weather conditions may be unsuitable. In such a case he will do best to take a magnetic bearing with a compass over his base line, and to calculate and plot his network on magnetic co-ordinates. If at a later stage he succeeds in measuring a sun azimuth he will be able to plot True North on his final map by setting out with a protractor the angle between Magnetic North and True North.

In the absence of a compass (although this is improbable if the survey has been planned properly) an assumed bearing can be used for purposes of calculation, and True or Magnetic North can be added later by setting out the appropriate angle from Assumed North.

Small triangulation schemes

If the final map is to be free of visible error, it must be based on a system of fixed points whose locations and heights are determined with an accuracy appropriate to the scale of the map. Most field workers will not be mapping on scales larger than 1:1000, and control points therefore need to be fixed accurately to within 250 mm. It is best to aim at an accuracy of at least 125 mm., to allow for any possible later work on a very large scale.

Station location and signals

A reconnaissance needs to be made on the ground for the precise location of fixed points. There is a difference between the requirements of national primary triangulation, for example, and the type of minor scheme described below. For a primary triangulation, the aim is maximum precision in the determination of station co-ordinates, and stations are therefore located with consideration for shape of triangle, intervisibility with other points without grazing rays which could cause refraction, and other factors affecting extreme precision. In a minor scheme, stations need to be located principally for their visibility from as much as possible of the area being surveyed. The highest points on two headlands enclosing a bay may be excellent for triangulation observation, but if these points are not visible within the bay itself, and it is the bay detail which is being mapped, then the two points are useless. Two lower points, possibly on cliff shoulders and visible over the whole of the bay, will be much more suitable. Visibility may be prejudiced, however, by background, and this should always be borne in mind. The station signal for observation throughout the survey is likely to be no better than a 3-metre striped pole held upright over the station mark by a tripod. The latter may be an old theodolite tripod or a small, folding metal brace. The former can be painted bright yellow or orange, which will help it to be distinguished against a woodland background, for example; the latter is much easier to install but much more difficult to see, whether against a confused background or against the sky, and some subsidiary signal may be necessary, such as an orange square propped against the signal base, which will be readily identifiable and can lead to easier intersection of the pole by an instrument sight.

As far as the terrain allows, fixed marks should be sited so that at least three are visible from every part of the area to be surveyed (not necessarily the same three from all parts, of course). If the country is rugged or the vegetation is thick or patchy, it may not be possible to meet this requirement, and marks must be sited to give as close a network as possible in order at a later stage to reduce the length of detail traverses between them. In open country with good visibility the number of points may not need to be very great. One example is given to illustrate the type of network laid down for surveys by first-year undergraduate parties. An ideal station mark is a large nail or screw inserted into a mushroom-shaped concrete pillar which has a splayed-out stem extending about a third of a metre below the surface. The hole is excavated, concrete poured in, rounded over the top and allowed to set: station identification marks can be inscribed on the smoothed wet surface with the end of a chaining arrow. Failing permanent marks of this type, stations can consist of a length of metal pipe of 25 mm diameter, hammered into the ground until only a short length shows above the surface, or, if the soil depth is very shallow, a 150 mm nail is tapped in until embedded with its head at surface level. The station mark should be surrounded with a small circle of stones laid loosely on the surface, to assist in its location

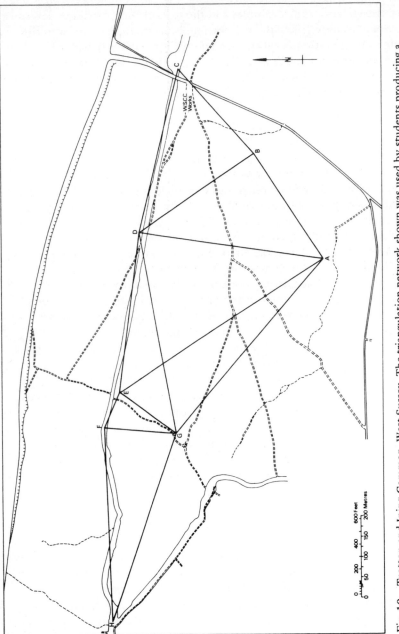

Fig. 10 Trotton and Iping Common, West Sussex. The triangulation network shown was used by students producing a vegetation map of the area on a scale of 1:1000.

each morning for the setting up of the station signal: if possible, a small cairn of stones may be raised over the mark if vegetation is likely to obscure it easily. A description and sketch must be made for each station mark, with bearings (usually magnetic bearings) to other stations or to readily identifiable objects, and measurements to nearby points of detail. The information must be adequate for easy re-location of the established mark, which may be very rapidly concealed by vegetation growth and movement. An unobtrusive but well-documented mark is preferable to a strikingly obvious marker in any area in which other people may be present. Survey pegs appear to have an irresistable attraction when unattended and are liable to be pulled up with great frequency. It is important that the hole can, if possible, be rediscovered from the station record. The surveyor must also be on guard against deliberate reinsertion of removed marks in positions other than the original, as the displacement may exceed the plottable error. Such reinsertion some distance from the calculated position can propagate errors and lead to great confusion if the false point is subsequently used for plane table resection or detail.

Care must be taken on the reconnaissance to avoid interruption of lines of visibility be obstructions such as tree branches. The observer standing at X may see Y clearly, looking over the top of a nearby tree or bush, but an observer at Y may be unable to see the base of the X signal. It is important to take all observations to station points to the base of the target pole, as it is not possible to guarantee the perfect verticality of this and sightings to the upper part of a pole may represent a plottable displacement from the station mark. For this reason reconnaissance sights from any station should not only be made from instrument height to ground level at the other stations, but should also be checked from ground level to instrument height at the other points, this may avoid later trouble and the possible necessity of returning to move the mark to another position with unobstructed lines of sight.

Base line

The most accurate determination of a base line length can be made using a modern electromagnetic instrument such as a geodimeter. Most departments will not have this type of instrument available on account of the high cost. For the purposes of this book therefore, it is assumed that the survey will be carried out without electronic equipment, or even catenary taping equipment, and that measurement along the ground surface will be required.

Two of the triangulation points should be so sited that the distance between them can be taped with ease. They should be separated by a reasonably even slope, with individual slope facets not exceeding 5°. Because all co-ordinate calculations will be derived from the calculation of the network, the base must be measured carefully. This will normally be done with a steel tape laid along the ground, each length being carefully aligned.

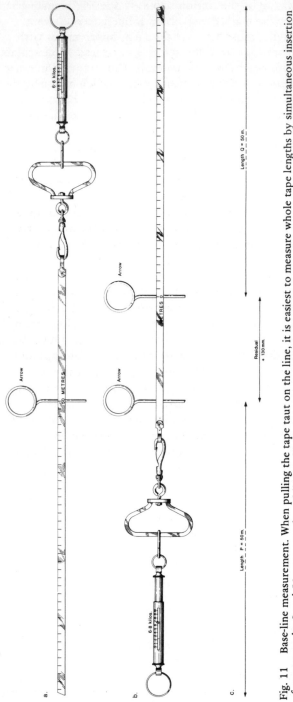

Fig. 11   Base-line measurement. When pulling the tape taut on the line, it is easiest to measure whole tape lengths by simultaneous insertion of arrows at the 0 and 50-metre marks, adding residual distances between tapes or subtracting overlaps. This gives a more accurate result than an attempt to hold the zero mark of one length exactly against the 50-metre arrow marking the end of the previous tape.

The tape should be pulled taut with a tension of about 7 kg (15 lbs), which can be checked with a spring balance hooked into the tape handle. If tape handlers find difficulty in holding the zero mark under tension against the station mark or against an arrow marking the end of the previous length, each handler should insert an arrow at a given call, one at the zero mark and the other at the 30-metre mark (or 50-metre mark, etc., depending on the length of the tape being used). The base length will be the total number of complete tape lengths + the final tape measurement ± the residual short distances between the arrows, measured with a scale (see fig. 11). All base line measurements should be made to the nearest 10 mm or better. The base should be measured at least twice, in opposite directions, with an Abney clinometer reading for slope being taken for each tape length or for each slope segment within the tape length. The two measurements of the base, corrected for slope (see table 3), should agree within 50 mm; if they do not, the work must be repeated until such agreement is obtained, when

Survey   *Telegraph Hill control net*
Date   *24 April 1972*
Observer   *A. B. Williams*

SLOPE	MEASURED LENGTH	SLOPE CORRECTION	HORIZONTAL LENGTH	RESIDUALS +	RESIDUALS −	NOTES
						Base AD to AE
					0·072	*Residual overlap*
E. 00° 50'	50·000	0·005	49·995			*1st tape*
				0·044		*Residual gap*
E. 01° 20'	50·000	0·015	49·985			*2nd tape*
				0·113		*Residual gap*
E. 01° 10'	50·000	0·010	49·990			*3rd tape*
					0·051	*Residual overlap*
D. 00° 20'	50·000	0·000	50·000			*4th tape*
					0·128	*Residual overlap*
E. 01° 10'	50·000	0·010	49·990			*5th tape*
				0·075		*Residual gap*
E. 02° 00'	50·000	0·030	49·970			*6th tape*
					0·137	*Residual overlap*
E. 03° 10'	13·800	0·021	13·779			*7th tape*
	313·800	0·091	313·709	0·232	0·388	
			+ 0·232			*+ gaps*
			313·941			
			− 0·388			*− overlaps*
			313·553			*Horizontal distance*

Fig. 12   Method of booking base measurement observations.

the mean value can be accepted. If the control net is extensive compared with the length of the base, greater accuracy is required, as any error in the base will affect all calculated lengths in the net. If the area being surveyed on a scale of 1:1000 were, for example, 1 km in length and the central base 100 m, the plottable error would be 250 mm and the two extremities of the survey should be accurate within 250 mm; in this case the error in the base

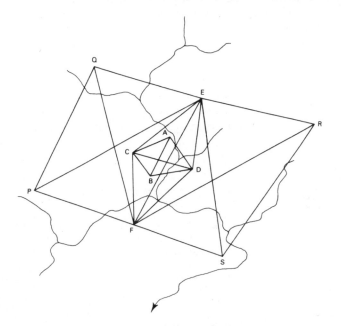

Fig. 13   A trigonometrical net across rugged country. The only location suitable for base measurement is *AB* on the valley floor, invisible from most of the area. Observation of all angles in the triangles *ABC*, *ABD*, *ACD* and *BCD* allows calculation of the length *CD*; similar observation of the triangles *CDE* and *CDF* allows calculation of the co-ordinates of *E* and *F* and so of the length of the side *EF* in the main triangulation.

should not exceed 25 mm. Figure 12 shows the method of booking base line measurements, by adding extra columns in a normal chaining book. The length of the base line depends on the total area to be surveyed. The base can usually be measured as one side of a triangle in the network, and in this case due consideration must be given to the shape of the triangles of which it forms a part: these triangles should not include angles of less than 30°.

   If it is not possible to measure one side in the triangulation network, it may be necessary to include a small base extension net. This may be forced upon the observer in some types of survey, for example in rugged country the only reasonably even stretch of land suitable for base measurement may be a short straight stretch of flood plain in a winding valley (fig. 13). In this case observation of all angles in the triangles of the base extension net enables calculation of one side of the main triangulation. It is clear that where

Observer	E. O. Davies			Date	17 July 1972			Survey	Telegraph Hill	

AT	TO	FACE	HORIZONTAL CIRCLE ° ' "	HORIZONTAL ANGLE ° ' "	VERTICAL CIRCLE ° ' "	VERTICAL ANGLE ° ' "	TIME	LEVEL E O	REMARKS
ΔD	B	L	01 17 21		00 21 43	+ 00 21 43		✓	
	A		25 37 31	24 20 11	01 38 10	+ 01 38 10		✓	
	C		54 44 57	53 27 37	01 11 15	+ 01 11 15		✓	
	F		255 23 23	254 06 03	357 17 52	− 02 42 08		✓	
	G		289 39 59	288 22 39	356 51 11	− 03 08 49		✓	
	B		01 17 19						
			01 17 20						
	B	R	181 17 20		179 38 18	+ 00 21 42		✓	
	A		205 37 32	24 20 11	178 21 46	+ 01 38 14		✓	
	C		234 45 01	53 27 40	178 48 45	+ 01 11 15		✓	
	F		75 23 25	254 06 04	182 41 59	− 02 41 59		✓	
	G		109 40 04	288 22 43	183 08 50	− 03 08 50		✓	
	B		181 17 23						
			181 17 21						
	B	L	45 00 42		00 21 42	+ 00 21 42		✓	
	A		69 20 53	24 20 12	01 38 09	+ 01 38 09		✓	
	C		98 28 20	53 27 39	01 11 17	+ 01 11 17		✓	
	F		299 06 48	254 06 07	357 17 53	− 02 42 07		✓	
	G		333 23 21	288 22 40	356 51 05	− 03 08 55		✓	
	B		45 00 39						
			45 00 41						
	B	R	225 00 46		179 38 10	+ 00 21 50		✓	
	A		249 20 58	24 20 15	178 21 48	+ 01 38 12		✓	
	C		278 28 18	53 27 35	178 48 46	+ 01 11 14		✓	
	F		119 06 50	254 06 07	182 42 00	− 02 42 00		✓	
	G		153 23 18	288 22 35	183 08 50	− 03 08 50		✓	
	B		225 00 40						
			225 00 43						

Fig. 14a   Field book entries for triangulation observations. Each round of horizontal angles starts and closes on the same point. Horizontal and vertical angles are calculated in the field as a check on the accuracy of observation.

Triangulation angles at ΔD

HORIZONTAL

BÂD	BÔC	BÔF	BÔG
24° 20' 11"	53° 27' 37"	254° 06' 03"	288° 22' 39"
20 11	27 40	06 04	22 43
20 12	27 39	06 07	22 40
20 15	27 35	06 02	22 35
24 20 12	53 27 38	254 06 05	288 22 39

	AÔC	AÔF	AÔG
	29 07 26	229 45 52	264 02 28
	07 29	45 53	02 32
	07 27	45 55	02 28
	07 20	45 52	02 20
	29 07 25	229 45 53	12" range RE-OBSERVE

		CÔF	CÔG
		200 38 26	234 55 02
		38 24	55 03
		38 28	55 01
		38 32	55 00
		200 38 27	234 55 01

			FÔG
			−34 16 36
			16 39
			16 33
			16 28
			11" range RE-OBSERVE

VERTICAL

DB	+ 00° 21' 43"
	21 42
	21 42
	21 50
	+ 00 21 44
DA	+ 01 38 10
	38 14
	38 09
	38 12
	+ 01 38 11
DC	+ 01 11 15
	11 15
	11 17
	11 14
	+ 01 11 15
DF	− 02 42 08
	41 59
	42 07
	42 00
	− 02 42 03
DG	− 03 08 49
	08 50
	08 55
	08 50
	− 03 08 51

Fig. 14b   Analysis of the angles observed at Station D and recorded in fig. 14a.

a base extension net is used, angles should not be less than 30° at any stage in the calculation to fix the co-ordinates of E and F. The length EF can then be calculated and the main triangulation computed.

Triangulation observation

The accuracy of the results must depend upon the accuracy of the angular measurement. If the co-ordinates of the fixed points are to be correct to within 125 mm, angles must be correct to within 15 seconds of arc at a range of about 1750 m, which is likely to be the maximum range for most surveys of this kind. If the observer has a theodolite reading to 1 second of arc, there is no difficulty in reaching the standard of accuracy required. Each angle can be measured in rounds on Face Left and Face Right. Each round may include as many network points as are visible, but must close back on the starting point. A page of an observation book is illustrated in fig. 14. The first and final readings to the starting point should agree within 15 seconds (or better, with a 1″ theodolite), the mean value of these is obtained, and this is subtracted from the circle reading to each other point, to obtain the corresponding angle. Face Left and Face Right values for each angle can then be compared, and meaned if they agree within 15 seconds. If they do not agree, the observer takes new Face Left and Face Right measurements of the angles in question. Vertical angles can be taken if it is desired to calculate heights from the theodolite observations, but heights are more accurately obtained by levelling.

Subtense measurement of multiple angles   If the observer has a theodolite reading only to the nearest 20 seconds, even with estimation to the nearest 10 seconds, accuracy becomes marginal. Repeated observations will not necessarily improve the situation, although the mean of three pairs of FL and FR observations may give slightly greater accuracy. A great increase in accuracy may be obtained by measuring a multiple of each angle, as for subtense measurement of distance (see p. 117). If the telescope, at P, is pointed first at X and then at Y, while the horizontal circle remains clamped, the correct angle XPY is set out on the horizontal circle, even though the micrometer or vernier may not be competent to measure it closer than to the nearest 20 seconds. For example, suppose the angle XPY is 47° 31′ 32″.1. The circle when pointing at X reads 229° 05′ 20″; when pointing at Y it reads 276° 37′ 00″. The angle is measured as 47° 31′ 40″. The small difference of 07″.9 cannot be detected with this instrument, which records the circle graduation nearest to the true value, in this case the next higher graduation. The angle actually set out on the horizontal circle is, of course, the true one not including this error of 07″.9. Repeat observations, even with different settings of the horizontal circle, cannot improve the answer, e.g. a pointing to X of 28° 11′ 00″ and to Y of 75° 42′ 40″ still gives an angle of 47° 31′ 40″ instead of 47° 31′ 32″.1. If, however, it were possible to set out an angle five times as large (i.e. 5 x 47° 31′ 32″.1 = 237° 37′ 40″.5) and the theodolite still read 28° 11′ 00″ when pointing at X, it would read 265° 48′ 40″ after turning through this angle, i.e. it would record an angle of 237° 37′ 40″. This divided by 5 would give 47° 31′ 32″, which is very close to the correct value.

It is possible to accumulate an angle 5 times (or more) the desired value, thus improving the accuracy beyond the direct capacity of the theodolite. The observer intersects X and reads the horizontal circle, e.g. 28° 11′ 00″; keeping the lower plate (i.e. the circle) clamped, he intersects Y and reads 75° 42′ 40″. He then keeps the *upper* plate clamped, unfastens the lower plate and intersects X a second time, completing the final intersection with the *lower* plate tangent screw. The instrument should still be reading 75° 42′ 40″, the circle having turned with the telescope. The lower plate is now left clamped, the *upper* plate is unfastened and Y is again intersected with the *upper* plate tangent screw. There is no need to read the micrometer or the vernier, but only to remember that a second movement of the telescope, again equal to the desired angle, has been made over the circle. As before, the upper plate is now kept clamped, the lower plate freed and X intersected for a third time, using the lower plate tangent screw. The lower plate is then again clamped, and the upper plate screws used to come on to Y for a third time. So the process continues, movements X to Y being made with the circle clamped and the upper plate moving over it, movements Y to X being made with the circle clamped to the upper plate and both moving together relative to the instrument base. A reading is taken after the fifth intersection of Y, giving in this example 265° 48′ 40″. Calculation is simple:

1st reading to X	28° 11′ 00″		
1st reading to Y	75° 42′ 40″	Approximate angle x 1 =	47° 31′ 40″
5th reading to Y	265° 48′ 40″	Approximate angle x 5 =	237° 37′ 40″
		÷ 5 = angle =	47° 31′ 32″

This method gives sufficiently high accuracy for angular measurement even with a 20″ instrument. The calculation of the first approximation acts as a check on correct counting of five measurements before the final reading is taken. With this technique there is, of course, no possibility of closing a round of angles on the starting point. Two dangers must be avoided: use of the wrong tangent screw in intersecting X or Y, and miscalculation of the multiple angle. If the single angle exceeds 72°, its fivefold aggregate will exceed 360°, e.g.

1st reading to X	173° 22′ 20″		
1st reading to Y	284° 17′ 40″	Approximate angle x 1 =	110° 55′ 20″
5th reading to Y	07° 59′ 20″	Approximate angle x 5 =	554° 37′ 00″
		÷ 5 = angle =	110° 55′ 24″

It is clear that the approximate angle x 5 cannot be 194° 37′ 00″. With multiples of larger angles it may be necessary to incorporate 720° or some other multiple of 360°. The one disadvantage of the method is that each angle at a station has to be measured separately instead of several angles being observed during a single round. Slowness is, however, compensated by accuracy. It must be stressed that it is preferable to use a 1″ theodolite if one needs greater accuracy than is possible with a 20″ instrument: the method described here is normally used professionally only with a subtense bar for optical measurement of distance (see p. 117).

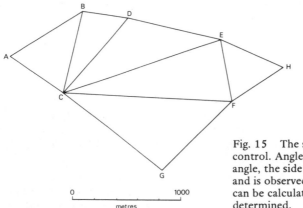

Fig. 15   The simplest form of triangulation control. Angles are observed in each triangle, the side *CD* is measured as a base and is observed for azimuth. All triangles can be calculated and station co-ordinates determined.

Adjustment of errors

When all required angles have been observed, the triangulation scheme must be computed. This is a relatively simple operation for the type of work envisaged. Unlike primary triangulation, the triangles may be regarded as plane and not spherical, and the refinements necessary for triangles with 70-kilometre sides do not apply to sides of 1 or 2 kilometres.

The simplest figure for inclusion in a triangulation scheme is a single triangle. A chain of simple triangles will suffice to provide basic control, although it is not the best method (see fig. 17 and the accompanying text (p. 26) on the importance of tie lines). If in the case illustrated in fig. 15 the side *CD* had been measured as a base and had been observed for azimuth, the triangles could be calculated in turn. Angles would be adjusted so that in $\triangle BCD$, for example, $\hat{B} + \hat{C} + \hat{D} = 180°$. (The triangle being small, we can assume that it is a plane triangle whose three angles total 180°. A large triangle, e.g. of side length 60 kms, in a national survey, would be calculated as a spherical triangle on the curved surface of the earth, and its three angles would total over 180°, the amount by which the total exceeds 180°, being known as the spherical excess.) If the three observed angles together differ from 180°, they are adjusted equally to total 180°. Examples might be as follows:

$$\hat{B} = \ 96° \ 21° \ 10''$$
$$\hat{C} = \ 26° \ 03' \ 55''$$
$$\hat{D} = \ 57° \ 35' \ 10''$$

Sum = 180° 00′ 15″   Correction −5″ per angle

or

$\hat{B} = 96° 21' 00''$
$\hat{C} = 26° 03' 45''$
$\hat{D} = 57° 35' 00''$
Sum = 179° 59' 45''   Correction +5″ per angle

or

$\hat{B} = 96° 21' 05''$
$\hat{C} = 26° 03' 45''$
$\hat{D} = 57° 35' 00''$
Sum = 179° 59' 50''   In this case the total correction is +10″,
with the largest correction (+4″) on the
largest angle at B, and corrections of +3″
on each of the other two angles.

When the angles have been adjusted, the sides are calculated from the formula

$$\frac{b}{\sin \hat{B}} = \frac{c}{\sin \hat{C}} = \frac{d}{\sin \hat{D}}$$

where b, c and d are the sides respectively opposite the angles $\hat{B}$, $\hat{C}$ and $\hat{D}$. If CD is measured, then

$$\frac{CD}{\sin \hat{B}} = \frac{c}{\sin \hat{C}} = \frac{d}{\sin \hat{D}}$$

and the other two sides can be calculated.

If C is used as the origin of the survey, it should be given false co-ordinates (if national co-ordinates are not known) such that every part of the area to be surveyed will have +Eastings and +Northings. This avoids negative co-ordinates at any stage in the calculations. The co-ordinates of D can be readily found, as fig. 16 shows. In the right-angled triangle XCD, $X\hat{C}D$ is known (the azimuth of D from C), CD has been measured, and $C\hat{X}D$ is 90°.

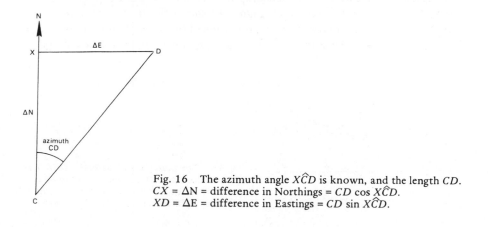

Fig. 16   The azimuth angle $X\hat{C}D$ is known, and the length CD.
$CX = \Delta N$ = difference in Northings = $CD \cos X\hat{C}D$.
$XD = \Delta E$ = difference in Eastings = $CD \sin X\hat{C}D$.

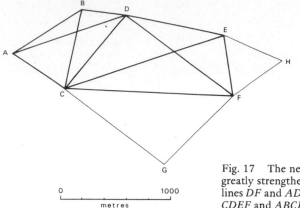

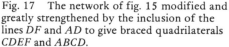

Fig. 17   The network of fig. 15 modified and greatly strengthened by the inclusion of the lines *DF* and *AD* to give braced quadrilaterals *CDEF* and *ABCD*.

Then

$$XD = \text{difference in Eastings} = CD \cdot \sin \hat{C}$$

and

$$CX = \text{difference in Northings} = CD \cdot \cos \hat{C}.$$

Reverting to fig. 15, the co-ordinates of *B* relative to *C* can be similarly calculated, and *B* relative to *D* will be calculated as a check on the two computations. Once the triangle *BCD* has been computed, adjacent triangles can also be calculated.

For many geographical surveys, this system of simple triangles will suffice, in that any errors in co-ordinates at the end of the chain may be too small to be plottable. If, however, the scale of the survey is to be 1:1000, for example, the errors carried forward from triangle to triangle may accumulate to the extent of a plottable displacement. To avoid this, any survey on a large scale should have a framework triangulation based on a more complex figure which can be adjusted as a whole. The ideal form of framework figure for most geographers will be the braced quadrilateral, easy to observe and to compute. Figure 17 shows the triangulation of fig. 15 modified and enormously strengthened by the addition of two lines, *AD* and *DF*, to give a system of two braced quadrilaterals and two single triangles.

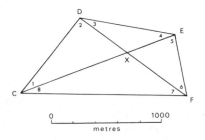

Fig. 18   The braced quadrilateral *CDEF*. *CD* is the measured base. The angles 1–8 are all observed. *X* is not an observed station, but merely the point of intersection of the diagonals.

Braced quadrilaterals   If the figure *CDEF* is considered in fig. 17, it is seen as a simple quadrilateral with two diagonals. Suppose that *CD* is the measured base, and that the angles 1 to 8 in fig. 18 have all been observed. The co-ordinates of *E* and *F* and the length *EF* could be calculated as in fig. 15 through the triangle *CDE* followed by the triangle *DEF*. They could equally well be calculated through the triangle *CDF* followed by the triangle *CEF*. It is obviously important that the results obtained should be identical irrespective of the route followed, i.e. the whole quadrilateral must be consistent within itself.

As a first step, angles must be adjusted for misclosure in each triangle. Using the angle number as shown in fig. 18, the following conditions must apply:

(1) $1 + 2 + 3 + 4 = 180°$ (angles of $\triangle CDE$)
(2) $3 + 4 + 5 + 6 = 180°$ (angles of $\triangle DEF$)
(3) $5 + 6 + 7 + 8 = 180°$ (angles of $\triangle CEF$)
(4) $1 + 2 + 7 + 8 = 180°$ (angles of $\triangle CDF$)

The interior angles of the quadrilateral *CDEF* must also total 360°, so

(5) $1 + 2 + 3 + 4 + 5 + 6 + 7 + 8 = 360°$

Also, from the triangles *CDX* and *EFX*,

$$D\hat{X}C = E\hat{X}F$$

and

$$1 + 2 + D\hat{X}C = E\hat{X}F + 5 + 6$$

Hence

(6) $1 + 2 = 5 + 6$

Similarly

(7) $3 + 4 = 7 + 8$

These seven condition equations are all interdependent. An alteration in angular values in any one equation must therefore affect all the others. Stage 1 of the calculation requires the satisfaction of these seven equations. It is best to start with equation (5) and to set out all eight angles, e.g.

Angle	Observed Value	1st correction (equation 5)	1st Adjusted Value
1 = DCE =	31° 11′ 25″	+2″	31° 11′ 27″
2 = CDF =	92 58 00	+3	92 58 03
3 = EDF =	25 20 25	+2	25 20 27
4 = DEC =	30 30 15	+2	30 30 17
5 = FEC =	81 03 35	+3	81 03 38
6 = DFE =	43 05 30	+3	43 05 33
7 = DFC =	34 44 00	+3	34 44 03
8 = ECF =	21 06 30	+2	21 06 32
	359° 59′ 40″	+20″	360° 00′ 00″

The 1st Correction is simply one-eighth of the total deficit or surplus. In this example the deficit is 20″, so that each angle should have a positive correction of 2″.5. In practice the four larger angles are each corrected by 3″ and the four smaller angles by 2″. Condition equation (5) is now satisfied. [Note: In the example given angles are rounded off to the nearest second throughout, but many surveyors prefer to calculate to tenths of a second and to round off to the nearest second at the end of the computation].

Subsequently it is sufficient to work with equations (6) and (7), as equations (1) to (4) can be obtained from (5), (6) and (7). For equation (6) we now have

Angle	1st Adjusted Value	Pair total	2nd Correction	2nd Adjusted Value
1	31° 11′ 27″	124° 09′ 30″	−4″	31° 11′ 23″
2	92 58 03		−5	92 57 58
5	81 03 38	124 09 11	+5	81 03 43
6	43 05 33		+5	43 05 38
	difference	19″	+1″	

and for equation (7) we have

3	25° 20′ 27″	55 50 44	−2″	25° 20′ 25″
4	30 30 17		−3	30 30 14
7	34 44 03	55 50 35	+2	34 44 05
8	21 06 32		+2	21 06 34
	difference	09″	−1″	

In the first case the difference between the pairs is 19″, and this is divided as nearly as possible into four equal parts. The three larger angles are each adjusted by 5″ and the smallest of the angles by 4″. Had the pair totals differed by 20″, it would have been possible to add 10″ to one pair and subtract 10″ from the other pair without altering the overall total of 360° 00′ 00″. As it is, the overall total has been increased by 1″, and the adjustment for equation (7) must therefore make a corresponding reduction by 1″.

In the second case the difference between the pairs is 09″. To restore the overall total to 360° 00′ 00″ we need adjustments of −5″ and +4″ and the larger of the two angles in the first pair is therefore given an adjustment of −3″. The overall total now remains at 360° 00′ 00″ and the angular equations are all satisfied.

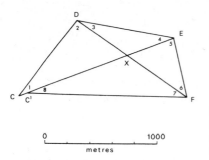

Fig. 19   Angle condition equations are satisfied, but C and C' do not coincide. A side equation is necessary to guarantee the coincidence of C and C'.

Nevertheless, the quadrilateral may be impossible of calculation even though the angular equations are satisfied. A calculation of lengths from CD in a clockwise direction through the three triangles CDE, DEF and EFC could result in C and C' not coinciding (fig. 19). A "side equation" is necessary to guarantee this coincidence, namely that $XC = XC'$, i.e. that $X\acute{C}/XC' = 1$.

This equation can be expanded:

$$1 = \frac{XC}{XC'}$$

$$= \frac{XC}{XC'} \cdot \frac{XD}{XD} \cdot \frac{XE}{XE} \cdot \frac{XF}{XF}$$

$$= \frac{XC}{XD} \cdot \frac{XD}{XE} \cdot \frac{XE}{XF} \cdot \frac{XF}{XC'}$$

In $\triangle CXD$, using the formula

$$\frac{a}{\sin A} = \frac{b}{\sin B} = \frac{c}{\sin C}$$

$$\frac{XC}{\sin CDX} = \frac{XD}{\sin DCX}$$

or

$$\frac{XC}{\sin 2} = \frac{XD}{\sin 1}$$

or

$$\frac{XC}{XD} = \frac{\sin 2}{\sin 1}$$

Similarly in $\triangle DXE$

$$\frac{XD}{\sin 4} = \frac{XE}{\sin 3}$$

or

$$\frac{XD}{XE} = \frac{\sin 4}{\sin 3} \quad \text{etc.}$$

Therefore we can devise an equation (8)

$$\frac{XC}{XD} \cdot \frac{XD}{XE} \cdot \frac{XE}{XF} \cdot \frac{XF}{XC'} = \frac{\sin 2}{\sin 1} \cdot \frac{\sin 4}{\sin 3} \cdot \frac{\sin 6}{\sin 5} \cdot \frac{\sin 8}{\sin 7}$$

Equation (8) must therefore be satisfied to ensure that C and C' coincide. This is called the "side equation", and is usually given in the form:

$$1 = \frac{\sin 2 \cdot \sin 4 \cdot \sin 6 \cdot \sin 8}{\sin 1 \cdot \sin 3 \cdot \sin 5 \cdot \sin 7}$$

If one does not have access to an electric calculator in the field, it is easier to add log sines than to multiply natural sines. Hence

log sin 2 + log sin 4 + log sin 6 + log sin 8

= log sin 1 + log sin 3 + log sin 5 + log sin 7

The sum of the logarithms of the sines of the "even" angles must equal the sum of the logarithms of the sines of the "odd" angles, i.e.

Σ log sin (even angles) = Σ log sin (odd angles)

The logarithms are taken from tables, together with the difference in each logarithm for each second of arc:

Angle	2nd Adjusted Value	log sin	Difference per second	
2	92° 57′ 58″	9.999 4178	−0001	(see below for
4	30 30 14	9.705 5189	0036	sign)
6	43 05 38	9.834 5452	0023	
8	21 06 34	9.556 4841	0055	
		Σ = 9.095 9660	Σ = 0113	
1	31 11 23	9.714 2237	0035	
3	25 20 25	9.631 4370	0044	
5	81 03 43	9.994 6940	0003	
7	34 44 05	9.755 7054	0030	
		Σ = 9.096 0601	Σ = 0112	
		Σ odd = 9.096 0601		
		Σ even = 9.095 9660		
		difference = 0941		

If there is no difference in the sum of the log sines, then the side equation is satisfied. Usually there is a difference which has to be removed by slight adjustment of the angles. In the example given, the difference is 0941 in the last four places of decimals. The sum of the log sines for the odd angles is greater than the sum for the even angles, so that the former needs to be reduced and the latter increased. As the angular equations have already been satisfied, any change in angular values at this stage must be equal and of opposite sign as between "odd" and "even" angles. (It can be seen from equations (1) to (7) that as each equation contains equal numbers of "odd" and "even" angles, any transfer of adjustment from "odd" to "even" will not affect the overall angular totals). A minimum change of 1″ increase on each "even" angle and of a corresponding decrease of 1″ on each "odd" angle would alter Σ even by +0113 and Σ odd by −0112 in the last four decimal places, thus reducing the overall differences by 0225. The total difference to be eliminated is 0941, so the required correction to each angle is 0941/ 0225 seconds = 4″.18. Each sine is therefore corrected by 4.18 × the difference for 1″, as follows:

Angle	2nd Adjusted Value log sine	3rd Correction	3rd Adjusted Value log sine	3rd Adjusted Value
2	9.999 4178	4.18 x −0001 = −0004	9.999 4174	92° 58′ 02″
4	9.705 5189	4.18 x +0036 = +0150	9.705 5339	30 30 18
6	9.834 5452	4.18 x +0023 = +0096	9.834 5548	43 05 42
8	9.556 4841	4.18 x +0055 = +0230	9.556 5071	21 06 38
			Σ = 9.096 0132	
1	9.714 2237	4.18 x −0035 = −0146	9.714 2091	31 11 19
3	9.631 4370	4.18 x −0044 = −0184	9.631 4186	25 20 21
5	9.994 6940	4.18 x −0003 = −0013	9.994 6927	81 03 39
7	9.755 7054	4.18 x −0030 = −0125	9.755 6929	34 44 01
			Σ = 9.096 0133	360 00 00

Always check the calculation by summing the odd and even log sines as before. In this example, the calculation emphasises the necessity for this, as one of the angles (angle 2) exceeds 90°. In such a case an increase in the angle gives a decrease in the sine, and vice versa, and so the positive or negative effect of a change must always be noted. In the example, angles 2, 4, 6 and 8 have all been increased by 4″ each, and whereas this has increased the log sine values for angles 4, 6 and 8, log sin 2 has changed in the opposite direction.

The quadrilateral is now fully adjusted for both side and angle conditions.

Calculation of quadrilateral   Co-ordinates of the triangulation points are now obtained by computation of the difference in Eastings and Northings over each side. As all angles are now fully adjusted, the same results should be obtained irrespective of the route followed in the calculation. For example, in the triangulation scheme illustrated in fig. 17, D may be taken as the starting point of the calculation. The co-ordinates of G can be obtained by calculating DE, then EF, then FG; or by calculating DC and then CG. Both routes are used, in order to provide the necessary check, as shown below.

Azimuth CD was observed as	$40°\ 27'\ 11''$	
Azimuth DC is therefore	$40°\ 27'\ 11'' + 180°\ 00'\ 00''$	$40°\ 27'\ 11''$
	$= 220°\ 27'\ 11''$	$+180°\ 00\ 00$
		$\overline{220\ \ 27\ \ 11}$
$E\hat{D}C = \hat{2} + \hat{3}$	$=\ \ 92°\ 58'\ 02'' +\ \ 25°\ 20'\ 21''$	
	$= 118°\ 18'\ 23''$	$92\ \ 58\ \ 02$
		$+25\ \ 20\ \ 21$
		$\overline{118\ \ 18\ \ 23}$
Azimuth DE	$=$ Azimuth $DC\ -\ E\hat{D}C$	
	$= 220°\ 27'\ 11'' - 118°\ 18'\ 23''$	$220\ \ 27\ \ 11$
	$= 102°\ 08'\ 48''$	$-118\ \ 18\ \ 23$
		$\overline{102\ \ 08\ \ 48}$
Azimuth ED	$= 102°\ 08'\ 48'' + 180°\ 00'\ 00''$	$+180\ \ 00\ \ 00$
	$= 282°\ 08'\ 48''$	$\overline{282\ \ 08\ \ 48}$
$F\hat{E}D = \hat{4} + \hat{5}$	$=\ \ 30°\ 30'\ 18'' +\ \ 81°\ 03'\ 39''$	$30\ \ 30\ \ 18$
	$= 111°\ 33'\ 57''$	$+81\ \ 03\ \ 39$
		$\overline{111\ \ 33\ \ 57}$
Azimuth EF	$=$ Azimuth $ED\ -\ F\hat{E}D$	
	$= 282°\ 08'\ 48'' - 111°\ 33'\ 57''$	$282\ \ 08\ \ 48$
	$= 170°\ 34'\ 51''$	$-111\ \ 33\ \ 57$
		$\overline{170\ \ 34\ \ 51}$
$D\hat{C}F = \hat{1} + \hat{8}$	$=\ \ 31°\ 11'\ 19'' +\ \ 21°\ 06'\ 38''$	
	$=\ \ 52°\ 17'\ 57''$	$31\ \ 11\ \ 19$
		$+21\ \ 06\ \ 38$
		$\overline{52\ \ 17\ \ 57}$
Azimuth CF	$=$ Azimuth $CD\ +\ D\hat{C}F$	$52\ \ 17\ \ 57$
	$=\ \ 40°\ 27'\ 11'' +\ \ 52°\ 17'\ 57''$	$+40\ \ 27\ \ 11$
	$=\ \ 92°\ 45'\ 08''$	$\overline{92\ \ 45\ \ 08}$

In $\triangle CFG$, adjusted values for a simple triangle were:

	$\hat{C} =\ \ 34°\ 52'\ 10''$	
	$\hat{F} =\ \ 48°\ 02'\ 38''$	
	$\hat{G} =\ \ 97°\ 05'\ 12''$	
Azimuth CG	$=$ Azimuth $CF\ +\ F\hat{C}G$	$92°\ 45'\ 08''$
	$=\ \ 92°\ 45'\ 08'' + 34°\ 52'\ 10''$	$+34\ \ 52\ \ 10$
	$= 127°\ 37'\ 18''$	$\overline{127\ \ 37\ \ 18}$
Azimuth FG	$=$ Azimuth $FC\ -\ G\hat{F}C$	$272\ \ 45\ \ 08$
	$= 272°\ 45'\ 08'' - 48°\ 02'\ 38''$	$-48\ \ 02\ \ 38$
	$= 224°\ 42'\ 30''$	$\overline{224\ \ 42\ \ 30}$

We now have to calculate the lengths of the sides through the triangles, using the formula

$$\frac{a}{\sin A} = \frac{b}{\sin B} = \frac{c}{\sin C}$$

In $\triangle CDE$
$$\frac{DE}{\sin 1} = \frac{CD}{\sin 4}$$

$$DE = CD \cdot \frac{\sin 1}{\sin 4}$$

CD was the measured base = 842.30 metres.

We have already obtained the logarithms of the sines of the angles.

	log CD	2.925 4668
	log sin 1	9.714 2091
	sum	2.639 6759
	−log sin 4	9.705 5339
DE = 859.30 metres	log DE	2.934 1420

In $\triangle DEF$ $EF = DE \cdot \dfrac{\sin 3}{\sin 6}$

	log DE	2.934 1420
	log sin 3	9.631 4186
	sum	2.565 5606
	−log sin 6	9.834 5548
EF = 538.28 metres	log EF	2.731 0058

Point D has been given co-ordinates of +3000.00 E +4000.00 N.

To obtain co-ordinates of E:

Difference in Eastings of E relative to $D = DE \cdot \sin$ (azimuth DE)
Difference in Northings of E relative to $D = DE \cdot \cos$ (azimuth DE)
Azimuth DE has been calculated as $102° \, 08' \, 48''$

log DE	2.934 1420	log DE	2.934 1420
log sin azimuth	9.990 1666	log cos azimuth	9.323 0765
log ΔEastings	2.924 3086	log ΔNorthings	2.257 2185
ΔEastings	+ 840.06	ΔNorthings	−180.81
Eastings of D	+3000.00	Northings of D	+4000.00
Eastings of E	+3840.06	Northings of E	+3819.19

Similarly to obtain the co-ordinates of F:

Difference in Eastings of F relative to $E = EF \cdot \sin$ (azimuth EF)
Difference in Northings of F relative to $E = EF \cdot \cos$ (azimuth EF)
Azimuth EF has been calculated as $170° \, 34' \, 51''$

log *EF*	2.731 0058	log *EF*	2.731 0058
log sin azimuth	9.213 9319	log cos azimuth	9.994 1048
log ΔEastings	1.944 9377	log ΔNorthings	2.725 1106
ΔEastings	+88.09	ΔNorthings	−531.02
Eastings of *E*	+3840.06	Northings of *E*	+3819.19
Eastings of *F*	+3928.15	Northings of *F*	+3288.17

To obtain the co-ordinates of *G*:

$$\text{In } \triangle CDF \qquad CF = CD \cdot \frac{\sin 2}{\sin 7}$$

The logarithms have already been calculated

	log *CD*	2.925 4668
	log sin 2	9.999 4174
	sum	2.924 8842
	−log sin 7	9.755 6929
CF = 1476.36 metres	log *CF*	3.169 1913

$$\text{In } \triangle CFG \qquad FG = CF \cdot \frac{\sin F\hat{C}G}{\sin C\hat{G}F}$$

	log *CF*	3.169 1913
	log sin $F\hat{C}G$	9.757 1776
	sum	2.926 3689
	−log sin $C\hat{G}F$	9.996 6695
FG = 850.55 metres	log *FG*	2.929 6994

Difference in Eastings of *G* relative to *F* = *FG* . sin (azimuth *FG*)
Difference in Northings of *G* relative to *F* = *FG* . cos (azimuth *FG*)
Azimuth *FG* has been calculated as 224° 42′ 30″

log *FG*	2.929 6994	log *FG*	2.929 6994
log sin azimuth	9.847 2629	log cos azimuth	9.851 6846
log ΔEastings	2.776 9623	log ΔNorthings	2.781 3840
ΔEastings	−598.36	ΔNorthings	−604.48
Eastings of *F*	+3928.15	Northings of *F*	+3288.17
Eastings of *G*	+3329.79	Northings of *G*	+2683.69

We have the co-ordinates of *G* calculated through *E* and *F*. Check by calculation through *C*.

Azimuth *DC* has been calculated as 220° 27′ 11″

log *DC*	2.925 4668	log *DC*	2.925 4668
log sin azimuth	9.812 1274	log cos azimuth	9.881 3492
log ΔEastings	2.737 5942	log ΔNorthings	2.806 8160
ΔEastings	−546.51	ΔNorthings	−640.94
Eastings of *D*	+3000.00	Northings of *D*	+4000.00
Eastings of *C*	+2453.49	Northings of *C*	+3359.06

In $\triangle CFG$ $CG = CF . \dfrac{\sin \hat{CFG}}{\sin \hat{CGF}}$

	log CF	3.169 1913
	log sin $\hat{CFG}$	9.871 3728
	sum	3.040 5641
	$-$log sin $\hat{CGF}$	9.996 6695
CG = 1106.36 metres	log CG	3.043 8946

Azimuth CG has been calculated as $127°\ 37'\ 18''$

log CG	3.043 8946	log CG	3.043 8946
log sin azimuth	9.898 7575	log cos azimuth	9.785 6463
log ΔEastings	2.942 6521	log ΔNorthings	2.829 5409
ΔEastings	+876.30	ΔNorthings	$-$675.37
Eastings of C	+2453.49	Northings of C	+3359.06
Eastings of G	+3329.79	Northings of G	+2683.69

Because this is not a completely rigorous method of adjustment, and the standard of observation is not meticulous, small differences may be found in the co-ordinates calculated through different routes. In the example worked, no difference occurs, but it would be adequate to accept a mean value of differing results provided that at the limits of the network the mean co-ordinates differ from each of the calculated values by less than the plottable error of the scale on which the survey is being made. In the chain shown in fig. 17, co-ordinates of A and B are calculated by two different routes, and co-ordinates of H are calculated from both E and F.

Final co-ordinates for the network can then be plotted on plane-table sheets for use in the field. This is often done by plotting on squared paper on the desired scale and then transferring the plotted points to the plane-table sheets either by the use of a tracing table, if available, or by the use of a sheet of tracing paper. The points should *not* be pricked through with a pin, which causes holes in the plane table sheet (see p. 158). Best results are obtained by drawing rectangular grid lines on the plane table sheet and plotting co-ordinated points directly on the gridded sheet, but if the grid lines are not drawn exactly at right angles errors will result. (Equally, of course, errors can be introduced during the first method while transferring from the master sheet to the plane table.)

Notes on calculation   Do not use angles of less than 20° unless this is quite unavoidable. The log sines of small angles change quickly with small angular changes, and so lengths calculated with the aid of small angles are liable to be in error.

It will have been seen that many logarithms are used more than once in the calculation. If the surveyor understands what he is doing he will readily recall which logarithms have already been used and will save time by using them again without repeated recourse to tables. It is not mathematically necessary to use antilogarithms for all values in the computation. The lengths DE, EF and FG, for example, have been found from their logarithms, but the calculation of the G co-ordinates could have been completed using only the logarithms on some of the intermediate stages. The value of having the lengths in metres is that these provide a check on the plotting of the network co-ordinates before these are transferred to the plane table sheets: scale distances between the plotted points are checked against the values obtained.

It is useful to calculate azimuths over all the network lines and to list these, for two reasons. The first is that these azimuths can also be used as a check on the plotting of the co-ordinates on the squared paper, using a protractor, but this is less accurate than the check of side lengths just described. The second is that theodolite traverses may be used at a later stage to provide a basis for detail mapping, and such traverses will start and finish on triangulation points, with starting and finishing angles observed to other triangulation stations. If the observer has listed azimuths with him he can check his angular misclosure in the field and can take repeat observations at once if these appear desirable.

Figures other than braced quadrilaterals   A polygon with a centre point is the other form of figure likely to be useful to the geographer. Such a figure may have three sides or more. Figure 20(a) shows the triangulation points already used, but grouped in two centre-point figures with additional observed triangulation stations at P and Q.

A quadrilateral with a centre-point, such as $ABCD$ and P, would be necessary if A and C were not intervisible, and B and D were also obscured from each other. The number of angle condition equations is clearly five:

$$
\begin{aligned}
(1) \quad & 1 + 2 && + 10 && = 180° \\
(2) \quad & \quad 3 + 4 && + 11 && = 180° \\
(3) \quad & \quad\quad 5 + 6 && + 12 && = 180° \\
(4) \quad & \quad\quad\quad 7 + 8 && + 13 = 180° \\
(5) \quad & && 10 + 11 + 12 + 13 = 360°
\end{aligned}
$$

Adjust (1) to (4) in turn, with one-third of the total correction being applied to each angle, and with odd seconds being applied to the largest or two largest angles. Then sum the adjusted angles at P, and correct to 360° by redistribution in equations (1)–(4). This may alter angles at P which were originally observed as a closed round to total 360°, but there is probably nothing to suggest that the observations at P were more careful or more accurate than those at A, B, C and D: all observed angles at the five stations of the figure are therefore equally susceptible to adjustment.

(a)

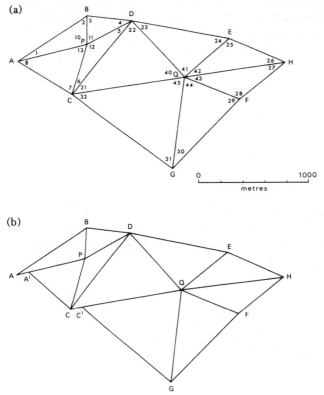

(b)

Fig. 20 (a) The triangulation points of fig. 17 grouped in a network of two figures with extra observed triangulation points at *P* and *Q*.
(b) There is still the necessity for side equations to ensure the coincidence of *A* with *A'* and of *C* with *C'*.

A side equation is still essential to guarantee that *A* and *A'* coincide. The equation is developed as in the case of the braced quadrilateral, i.e.

$$\frac{PA}{PA'} = 1$$

or

$$1 = \frac{PA \cdot PB \cdot PD \cdot PC}{PB \cdot PD \cdot PC \cdot PA'}$$

(6) $$1 = \frac{\sin 2 \cdot \sin 4 \cdot \sin 6 \cdot \sin 8}{\sin 1 \cdot \sin 3 \cdot \sin 5 \cdot \sin 7}$$

The adjustment to satisfy this equation is made as before.

Note: The angles at *P* do not enter into this equation, so sin 11, sin 12, etc. are not involved. Satisfaction of the side equation does not alter the centre-point angles at *P*.

Similarly, in the figure *CDEHFG*, with centre-point Q, there will be a condition equation for each triangle, plus one for the centre-point, plus the side equation to guarantee coincidence of C and C'.

$$
\begin{aligned}
21 + 22 + 40 \qquad\qquad\qquad\qquad &= 180° \\
23 + 24 \quad\ + 41 \qquad\qquad\qquad &= 180° \\
25 + 26 \qquad\ + 42 \qquad\qquad &= 180° \\
27 + 28 \qquad\qquad\ + 43 \qquad &= 180° \\
29 + 30 \qquad\qquad\qquad + 44 &= 180° \\
31 + 32 \qquad\qquad\qquad\qquad + 45 &= 180° \\
40 + 41 + 42 + 43 + 44 + 45 &= 360°
\end{aligned}
$$

$$
\frac{\sin 22 \cdot \sin 24 \cdot \sin 26 \cdot \sin 28 \cdot \sin 30 \cdot \sin 32}{\sin 21 \cdot \sin 23 \cdot \sin 25 \cdot \sin 27 \cdot \sin 29 \cdot \sin 31} = 1
$$

Adjustment and calculation is as before.

Any centre-point figure of n sides will therefore have $n + 2$ condition equations. Once adjustment is complete, co-ordinates are calculated both ways round the figure to provide a check.

Satellite stations

It is sometimes convenient to use as a fixed mark a prominent feature such as a rock pinnacle which can be clearly seen from a large part of the area being surveyed, but which cannot readily be used as an observing station.

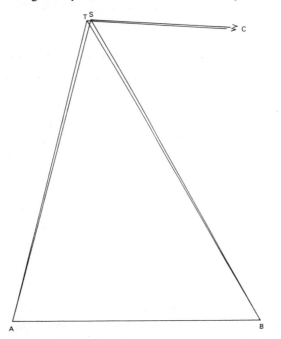

Fig. 21   Observations from A, B and C are made to the prominent feature T. Observations to A, B and C are made at the satellite station S, as T cannot be occupied.

In such a case it is possible to observe angles to the station from other points, but it is impossible to complete the network with observations at the point. If, however, a theodolite can conveniently be used only a short distance away, and observations can be made from this satellite position, it is possible to calculate the angles at the point required.

Figure 21 illustrates observations from A, B and C to the triangulation point T. Observations to A, B and C are made from the satellite station S. The angles measured at S are to be adjusted to give the correct values for angles at T, a process known as reduction to centre. The distance ST must be small (not more than $\frac{1}{40}$ th the length of any ray from S to another observed station).

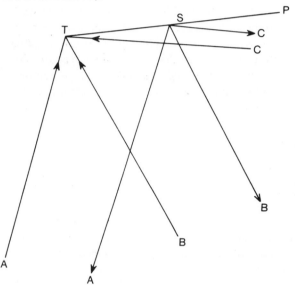

Fig. 22   An enlargement of part of fig. 21, with the addition of the line TS produced to P.

Figure 22 illustrates the satellite position on a larger scale. The line TS can be produced to any point P.
Required to find $B\hat{T}A$.

$$P\hat{S}A + A\hat{S}T = 180° = A\hat{S}T + S\hat{T}A + T\hat{A}S$$

(1) $P\hat{S}A$ $=$ $P\hat{T}A + T\hat{A}S$

Similarly

(2) $P\hat{S}B$ $=$ $P\hat{T}B + T\hat{B}S$

Subtracting (2) from (1)

$$P\hat{S}A - P\hat{S}B = P\hat{T}A - P\hat{T}B + T\hat{A}S - T\hat{B}S$$

$$B\hat{S}A =\qquad B\hat{T}A + T\hat{A}S - T\hat{B}S$$

or

(3) $B\hat{T}A$ = $B\hat{S}A - T\hat{A}S + T\hat{B}S$

In the very narrow triangle ATS

$$\frac{ST}{\sin T\hat{A}S} = \frac{AT}{\sin A\hat{S}T}$$

whence

(4) $\sin T\hat{A}S$ = $\sin A\hat{S}T \cdot \dfrac{ST}{AT}$

The point T has been included in all rounds of angles at S, so that $A\hat{S}T$ is known. It may be possible to measure ST directly with a tape, correcting for slope, but probably this is impossible: if T could be readily occupied, there would be no need for a satellite station. The distance ST will probably have to be calculated, setting out a small triangle QST as in fig. 23. QS can be measured and corrected for slope. The angles $S\hat{Q}T$ and $T\hat{S}Q$ are measured with the theodolite, and $Q\hat{T}S$ obtained by calculation.

$$Q\hat{T}S = 180° - S\hat{Q}T - T\hat{S}Q$$

Then

$$\frac{ST}{\sin S\hat{Q}T} = \frac{QS}{\sin Q\hat{T}S}$$

Provided that the triangle QST is well-conditioned, i.e. as near equilateral as is possible, with no angle less than 30°, the measurement obtained for ST will be sufficiently accurate. Because this length ST affects later cal-

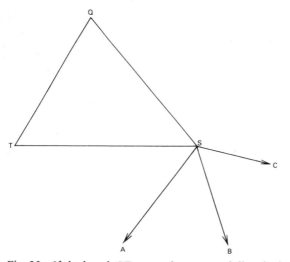

Fig. 23   If the length ST cannot be measured directly, it can be calculated by setting out a small triangle STQ, measuring QS and the angles at Q and S.

culations, the length QS should be the mean of two or four measurements with an accuracy of 10 mm, and the angles $S\hat{Q}T$ and $T\hat{S}Q$ should be checked with multiple-angle observations as described on page 22.

In $\triangle ATB$, $T\hat{A}B$ and $T\hat{B}A$ have been observed.

Assume

$$B\hat{T}A = 180° - T\hat{A}B - T\hat{B}A$$

$$\frac{AT}{\sin A\hat{B}T} = \frac{AB}{\sin B\hat{T}A}$$

AB is known, so AT can be calculated.

Equation (4) can now be calculated for $T\hat{A}S$. A similar calculation will give $T\hat{B}S$, so that equation (3) can be solved for the angle at T.

If observations are being made with a 1″ theodolite and angles are being adjusted to fractions of a second, the calculations of the very small angles at A and B may require a special form of computation in view of the rapid changes in the value of the sine with changes in a small angle. In this case calculate equation (4) in the form:

log sin $A\hat{T}S$		log AT (m)	
log ST (m)	_____	log sin 1″ of arc	4.685 5749
Sum (i)		Sum (ii)	
Sum (ii)	_____		
Subtract =		= log $T\hat{A}S$ in seconds of arc	

For calculations to the nearest second, however, this is unnecessary.

If the satellite is used as set out above, a check on the calculated small angles should be made by observing the angles $T\hat{A}S$ and $T\hat{B}S$ with the multiple-angle technique. However, whenever possible avoid the use of satellite station corrections. If it is unavoidable, it is usually easier for the non-specialist to observe to S as well as from S, thus obtaining co-ordinates for S in the ordinary way. If T is observed from S, the azimuth ST can be readily obtained. If the length ST is measured directly or indirectly, the difference of the co-ordinates between S and T can be computed, and the co-ordinates of T calculated and plotted as a useful landmark instead of as a network station.

Heights

Trigonometrical observations

With a small scheme of the size envisaged it should be possible to use a level in order to find accurate heights for all the triangulation stations. Two days of work by an observer with one helper should be adequate for this purpose. Alternatively, if the height of one station is known, vertical angles can be measured to the others and the heights calculated. The second method adds little to the overall time required, but is less accurate. On the

other hand, an observer required to work single-handed could observe trigonometrical angles without assistance.

Observation and calculation   The observation of vertical angles is carried
out in much the same manner as azimuth observations, but without the need
for use of the dark filter and without the pressure of time. After completion
of horizontal angles at a station, the observer takes one Face Left and one
Face Right reading to each of the other stations visible. Before each intersection he should make certain that his sensitive spirit level is central, thus
avoiding corrections or errors, and he should then bring the crosswire on
to the target pole at a height above the ground equal to the height of the
telescope above the station mark. If he is using target poles of identical
marking, he has only to hold a station pole alongside the theodolite to see
the mark at which he should aim on similar poles elsewhere. He should be
able to estimate the corresponding positon on other poles to within 20 mm.
Face Left and Face Right angles should agree within 30″ of arc. It is important to ensure that the Face is correctly entered in the field book in
order that there should be no doubt whether the vertical angle is one of
elevation or depression. The mean of the two readings is accepted as the
vertical angle. For lines not exceeding 2 km in length it is usually sufficient
to calculate the difference in height between the observer's station and the
target station by multiplying the horizontal distance (obtained from the
network calculation) by the vertical angle (the mean of the readings from
each end). To this can then be applied a correction (see fig. 25), which gives
an approximate total compensation for refraction and for curvature of the
earth for distances up to 5000 m. It can be seen from the formula that even
on short lines this correction is appreciable if one is aiming at accurate results,
as the target station will always appear to be lower than is actually the case,
and the correction is always added to the calculated height. In fig. 24 a
horizontal sight from the instrument at A to a target pole at B meets the
pole at height Y and not at X. For observation of vertical angles the observer at A would sight to point X and would thus obtain an angle of
depression although A and B are at the same height.

If trigonometrical heights have to be used, it is highly desirable to level
to a station in the centre of the area and to calculate the trigonometrical
heights outward from this point. As with co-ordinates, calculations should
be carried through a network by different routes in order to check both the
calculations and the observations. The values finally accepted represent a

Fig. 24   Effect of earth curvature on trigonometrical sights. The horizontal sight from
the instrument at A meets the target pole at Y. If the observer sights on X, the same
height above B as is the telescope above A, he will obtain an angle of depression although
the points A and B are at the same altitude.

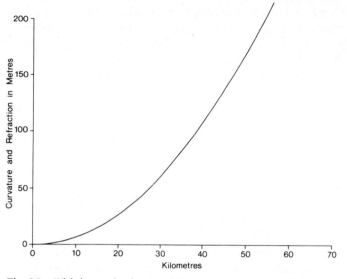

Fig. 25   With increasing length of line of sight, the altitude error XY of fig. 24 grows rapidly, further increased as the line of sight rises higher into the atmosphere and is refracted as in fig. 7 (p. 10). The graph shows the total effect against the length of line, approximately equal to $(0.5 - k)\,\dfrac{d^2}{R}$ metres

where k = the coefficient of refraction (about 0.07)
   d = length of line in metres
   R = mean radius of earth in metres (about 6369777).

compromise in this class of work. In the network in fig. 17, a height for D calculated from C would have more weight than a value reached by calculation from C via F and E. Values obtained over short lines are likely to be more reliable than those obtained from observations over long lines because of the increasing magnitude of the corrections with distance, as seen in fig. 25. This graph should make it clear why the non-specialist should avoid height calculation over lines of any considerable length and should, in particular, avoid attempting to bring a height value into his survey by a long-range observation to a known point in a national survey.

Heights by levelling

Much more accurate results can be obtained by using a level to determine heights of the network points. Bench mark lists for the area can be obtained for a small fee from the Ordnance Survey in the United Kingdom, and the most convenient marks selected for use. It should be remembered that even the most up-to-date published lists may be found inaccurate in the field due to recent building operations or public works, and any programme of work should allow adequate time for the substitution of a more distant bench mark for that originally selected for use.

   Detailed procedures for levelling are described on pp. 125–8. In fixing triangulation station heights it is necessary to work on closed circuits, i.e.

to start and finish on bench marks (although not necessarily the same mark) so that any accumulated error can be detected and adjusted. The misclosure should not exceed about 15 mm in 500 m, 30 mm at 1500 m, or 60 mm at 6000 m, with an experienced observer with a good instrument working over reasonable terrain. A non-specialist using a quickset level may be content with lower standards. Any misclosure is assumed to have occurred uniformly throughout the circuit, and correction is distributed evenly.

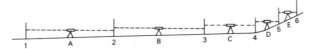

Fig. 26   It is important that the fore and back sights from the level telescope are of equal length, in order to reduce possible collimation errors. $1A = A2$, $3C = C4$, $5E = E6$, etc. Lines of sight inevitably are shorter on steep slopes.

If good results are to be obtained, certain precautions need to be taken. The lengths of foresight and backsight from each instrument position should be approximately equal. This is easily maintained if the staff man counts his paces as he moves forward to a new position, so that paces from the last staff point to the instrument equal those from the instrument to the new staff position. When the observer moves forward with the level, he should site its new position with some consideration of terrain, so that the staff man shall have the hope of achieving equality of distance. It is more important to have equal fore and back sights from the level than it is to have equality of sights on to any one staff position. In fig. 26, $1A = A2$, $2B = B3$, $3C = C4$, $4D = D5$, etc. As the slope becomes steeper, the possible length of line decreases. There is no sense in trying to make $3C = B3$, because if $3C$ were to equal $C4$, 4 would be well up the hillside above the horizontal line of sight.

Fig. 27 XY represents the earth's surface, points X, P, Z and Y being all at the same altitude.
GD represents the level surface.
HN represents the plane of the horizontal at P.
$XG = DY$ = height of telescope crosswire above ground at P. The horizontal line of sight at P gives staff readings H and N, but as $XP = PY$, $XH = YN$ to give no difference in height between X and Y, the curvature errors GH and DN cancelling out. If there is also an instrumental error of collimation, so that the line of sight is not horizontal when the bubble of the instrumental spirit level is in centre, the line of sight will be PA not PH, and PB not PN: but the errors AH and BN are equal, and will also cancel each other out, provided that P is midway between X and Y.
   If the level is not midway between staff positions, as is the case with X and Z, errors are introduced. Curvature error GH is not balanced by LO, nor collimation error AH by CO.

Lines of sight should not be too long. The errors due to refraction and curvature of the earth, mentioned above in connection with trigonometrical height observations, apply with equal force to levelling sights. In general, sights should not exceed about 50 m using a quickset level, partly to avoid

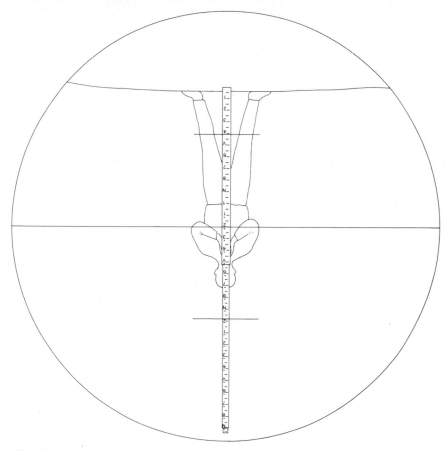

Fig. 28   The image of the levelling staff is usually inverted in the telescope view. The upper stadia reads 1.97 m, the lower stadia 0.41 m. The difference is 1.57, which indicates a distance from the observer of 157 m. As a check, the main crosswire reads 1.20 m: lower stadia to crosswire = 0.79, and crosswire to upper stadia = 0.78.

these errors and partly because reading of staff graduations becomes increasingly difficult as distance increases. A more powerful telescope may allow readings over greater distances, and this is in order provided that equality of sight length is preserved. If backsight and foresight are equal, then errors due to curvature and refraction cancel out. Figure 27 shows the balancing of curvature and other errors, and stresses the need for equality of distance. Pacing may be inaccurate on rough ground, so an additional check is desirable. This is easily made by reading the upper and lower stadia wires in the telescope, subtracting the smaller from the larger and multiplying by 100. Figure 28 shows stadia readings for distance measurement to the nearest metre. If it is found that the foresight and backsight distances differ by more than a few metres, the forward staff position is changed to give greater approximation to equality. The optical measurement of distance by multiplying the intercept (Upper Stadia reading − Lower Stadia reading) by 100 is known as tacheometry, and is considered in more detail on pages 67–9.

Stadia readings should always be taken during levelling operations. In practice, heights can be calculated without stadia readings, but these act as checks on the accurate reading of the horizontal crosswire. The interval between the lower stadia and the horizontal crosswire should equal that between the crosswire and the upper stadia: there may be a difference of 0.01 m, as in fig. 28, but with any larger discrepancy a check should be made of all three readings. Stadia readings thus immediately identify errors of observation which, if not at once detected, could invalidate all the work in a levelling circuit.

Although the primary purpose of a levelling circuit is to produce heights for the triangulation points, it is obviously advantageous to mark along the line of working some points which can be used at later stages in the survey. It is sufficient to hammer into the ground coloured wooden pegs, before the work is done, so that the staff may be held on the peg tops and the heights of these determined. Such pegs can be used as temporary local bench marks for subsequent contouring, and can be fixed for position if wished.

The careful heighting of the fixed and temporary marks by the use of a level is important. A geographer may require contours to be mapped with relatively small vertical intervals, such as 1 m. This is a greater standard of accuracy than is necessary for most topographical mapping.

The heights of all fixed marks should be listed and taken to the field during fixation of detail, as they will be used at some stage in the mapping.

2 The fixation of minor points

The nature of the terrain will determine the extent to which minor fixed points are necessary. In open country of rolling character, with good visibility, the triangulation points as shown in fig. 17 might be sufficient for the subsequent mapping of detail by plane table. This detail could be fixed by resection from the triangulation stations at almost any point in the area. However, if the area were rugged, minor points might be necessary for mapping in the valleys from which the triangulation stations were invisible, or if the vegetation included scattered woodland, shrubs or high bracken, for example, the triangulation stations might well be invisible over a large part of the area. In either of these cases it would be necessary to fix additional points from which mapping could continue. Such fixation of minor points is best done by *traverses*, a traverse being a surveyed line or route. The type of traverse used depends upon the degree of accuracy required for the final map. The most accurate form is the theodolite traverse and the least accurate is the compass traverse. Plane-table traverses form a useful mid-way technique, quicker and easier than the theodolite traverse, and far superior to the compass traverse.

Theodolite traverses

Figure 29 shows the previous network of triangulation points with a series of traverses either between these points or starting and ending on the same point. The traverse network shown is fairly dense, and implies close detail and obscured views. Each of the traverse stations is marked with a brightly painted wooden peg hammered into the ground, or with a nail hammered into bare rock. The traverse station is the centre of the nail head or the centre of a small hole in the peg top—the hole is easily made with the point of a metal chaining arrow.

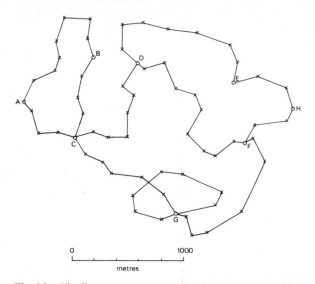

Fig. 29   The lines represent traverses between the stations of the minor triangulation scheme shown in fig. 17. Each cross indicates a traverse station marked on the ground. For a detailed survey of fairly thick vegetation a dense system of fixed points is necessary for detail mapping: in open areas of grass or heather with good visibility and less plant variety it is possible for fixed points to be spaced less frequently.

A reconnaissance should be made for each traverse, for the siting of the pegs. The peg positions are chosen for

(a) convenience of the traverse
(b) convenience of observing
(c) usefulness for subsequent work, i.e. with regard to visibility around them.

Inevitably some traverse points have little subsequent use, but are essential for the traverse itself. A traverse line passing through woodland should be surveyed with the minimum disturbance of the vegetation, and may require very careful siting of pegs to allow intervisibility between trees, and to allow chaining of distances from peg to peg. The convenience of the observer must always be remembered also: a nail on a rock pinnacle may be admirable for (a) and (c) above, but may not have room for the theodolite tripod and the observer. Mark the peg sites clearly, making detailed notes on their sites, so that they are readily found during the observation of the traverse. Heather pushed aside during the hammering of a peg may well return to its original position in the course of a few hours and may completely cover the peg. If, during the traverse observation, a substitute peg is sited in the approximate locality because the original peg cannot be found, confusion may later arise if the first peg is rediscovered. Each peg should be clearly marked with a code letter or number at the time of its siting: this can be done with a hard pencil or a ball-point pen *after* the peg has been hammered in. Make a count of paces from peg to peg during the reconnaissance. This has three virtues:

(i) it provides an approximate value for distance, and so assists subsequent finding of pegs;

(ii) it acts as a check against gross error in subsequent chaining;

(iii) because it requires more or less direct walking from peg to peg, it reveals any difficulties likely to be encountered in the chaining, e.g. patches of deep bog which are almost impassable and which are not readily recognisable from the traverse pegs between which they are situated: if the pegs are moved before observations start the traverse lines may then avoid the obstacle. Often a move of no more than 5 m will achieve this improvement.

The surveying of the traverse requires observation of the angles and measurement of the lines, with offset measurements to detail close to the traverse line. With adequate labour, both operations proceed simultaneously, a competent assistant supervising the chaining while the observer reads the angles. The geographer mapping for research purposes will not usually have adequate labour, and the two operations should therefore be performed separately. If this is done, a theodolite traverse can be carried through with only two unskilled assistants.

Measurement of distances

Instruments   The measurement of distances is known as chaining, even though it is now usually performed with a metal tape. Depending on the graduations on the tape, readings may be taken to the nearest 10 mm or 25 mm, estimating as necessary. The steel tape is stored on a circular holder from which it is unwound at the start of the day. Brass handles with swivel joints are attached at each end. The tape must always be treated with care, as it is very easily broken if it is bent sharply: repair requires the fitting of a metal sleeve across the broken ends, with rivetting or welding. The tape may be of any length from 30 m to 300 m. For most field work the shorter lengths are to be preferred, as they need less labour in handling. The zero and end marks will be a short distance from the handles. Methods of marking tapes vary. The best have brass tags engraved with numbers set at 10-metre intervals, with brass rivets, circular or square, to denote metres and tenths. Some tapes have graduations printed on one side, but these tend to be less satisfactory, as the marks become indistinct with time. At the end of the day, as the tape is slowly rewound on its reel, it should be passed through an oily cloth and wiped clean. If it is raining, the tape should be cleaned and dried under cover as soon as possible to avoid rusting, horizontal tests with a spring balance in order to be able approximately to gauge the feeling of this standard tension, and should endeavour to take all tape measurements with this degree of strain.

Accessories used with the tape are arrows, linen tape, ranging rods and spring balance. *Arrows* are straight lengths of wire about one-third of a metre in length, sharpened at one end and looped at the other. They are used for recording the number of tape lengths over a measured line by marking the end position of each tape length. They are issued in sets of 10. Before use in the field, a piece of coloured tape, preferably red or white, should be tied through each loop to render it more visible in grass or other vegetation. The *linen tape* is graduated in metres and fractions of a metre, and is wound in a circular case. It is used for short offset measurements at right angles to the measured line. It must never be allowed to get twisted in its case, when it is liable to break during unwinding: it should always be wound in through two fingers held across the aperture of the case, to ensure that twisting does not occur. It must never be wound in a dirty condition, as grit or mud inside the case can cause jamming: if work has to be done in wet or dirty conditions, either the tape must be wiped clean each time before winding, or, better, it must be carried around unwound if many measurements are being taken, cleaning being carried out before the final winding at the end of the day. If wet, the tap should be pulled out to its full extent on return to base, and allowed to dry out. A tape swells when wet, so that the case may not accommodate the whole length, and if force is used to wind it all in it is highly probable that the force required later to unwind it will be sufficient to snap the tape inside the case.

Ranging rods are wooden or aluminium poles usually about 2 m in length, with a pointed metal shoe at one end, and painted in sections of red and white, or of red, white and black. Each painted section is 200 or 250 mm in length, so that a rod can be used for measuring short lengths or heights within the limits of accuracy of estimation of subdivisions of the sections. Longer poles are used for special purposes, e.g. 3-metre rods may be used with target tripods to mark triangulation stations. Rods can be acquired in interlocking sections of one-metre length, which are particularly useful if they have to be carried in a small vehicle, on a bicycle, or on the back over an appreciable distance. It is now possible to buy miniature, lightweight, adjustable tripod supports (fig. 30) which will hold a rod upright. With the assistance of these it is possible for a theodolite traverse to be executed by the observer with only one assistant, although this involves the assistant in a considerable amount of extra walking to and fro.

The *spring balance* is a normal cylindrical balance with at one end a hook for gripping the brass handle of a tape, and at the other end a wire loop for holding by hand. It is used to ensure a standard tension on the tape, in British practice usually 15 lbs (6.8 kg), but for minor survey purposes it is usually unnecessary except perhaps for base measurement (see p. 18). Nevertheless, the surveyor and his assistants should make a number of

which can rapidly destroy printed graduation marks and can make the tape brittle and more liable to fracture. The tape should be checked periodically against a fixed base or a standard tape.

Fig. 30   Ranging rod in miniature supporting tripod.

If the survey is being done for any reason in non-metric units, a 100-foot tape may be used, or if the standard of accuracy required is not too high, a 100-foot chain may be used instead of a steel tape. The chain consists of

100 links, each being a length of strong wire bent at the ends, connecting with each of its neighbours through small oval rings. Brass handles are fitted at each end, with a swivel joint to prevent twist. The first and last links are shorter, so that the length from the outer edge of the handle to the first join is equal to one link: the overall length of 100 ft is from the outside edge of one handle to the outside edge of the other. Every tenth link has at the join a distinctive brass tag as shown. The chain can withstand rougher treatment than the steel tape, and is easily read to the nearest foot (or by estimation to the nearest half-foot), but the rings tend to open with time, thus altering the length of the chain, it is heavier than the tape and it collects mud, grass, etc., much more readily and it is difficult to clean. It is

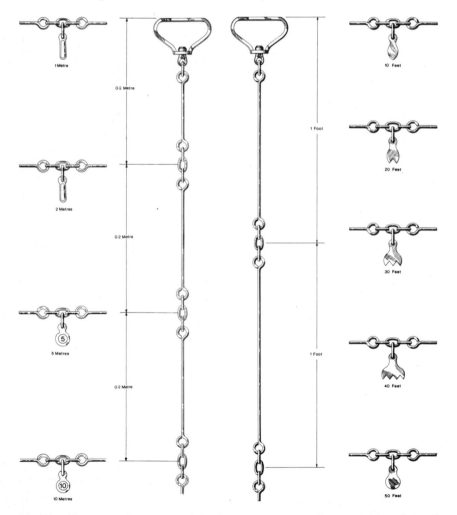

Fig. 31   For many minor surveys it is adequate to measure distances with a chain. The illustration shows the distinguishing tags of the 100-foot and of the 20-metre chains.

stored by laying out 0 to 50 ft, with 50 to 100 laid alongside so that the two handles are together. Links 50 and 51 (i.e., one each side of the 50-foot tag) are placed together in one hand, and each successive pair of links is folded in above, until the two end links, with handles, complete the operation, when a strap is fastened round the centre of the pack. To unfold the strap is removed, the two handles are held firmly in one hand and the chain is thrown upwards and outwards with the other hand; ideally it extends to 50 ft, when the two halves of the chain are separated and work can begin. Failure to pack the chain correctly, or to unfold it adequately, can lead to a prolonged period of unravelling a tangled skein. The most useful metre chain is 20 m in length. It has links of 0.2 m and tags for metre readings as shown in fig. 31.

Measurement procedure   The measurement of a line is considerably facilitated if certain routines are followed. One man is required on each end of the tape, the more experienced being the "follower" on the rear end, as he has the responsibility for aligning the leader. The leader goes forward along the line, holding the front end of the steel tape, 10 arrows and a ranging rod, until the whole length of the tape has been taken out. If the leader counts his paces as he goes, he will know when he has reached the required distance: this avoids jerking the tape and shouting by the follower. The leader then stands sideways to the follower, so that the latter's view is not obstructed, and holds the ranging rod so that it hangs vertically from his hand. The follower usually also carries a ranging rod and holds this vertical on the mark at which he stands. Alignment of the leader's pole is then between the follower's pole and the target pole. The follower signals to the leader to move the pole until it is on line, when the leader places the foot of the pole on the ground. (It avoids confusion if the follower signals with his left hand for a move to the left, and with his right hand for a move to the right: sweeping gestures for a large move, less generous gestures for a small move. When the leader's pole is on line, the follower should hold a hand above his head.) The follower then lays aside his pole, and concentrates on holding the zero of the tape against his mark. The leader pulls the tape taut alongside his pole, and on a call from the follower also lays aside his pole and inserts an arrow alongside the 30-metre or other terminal mark.

The process is then repeated for the next length, except that the leader leaves in the ground the arrow he has inserted. The follower carries his ranging rod, a field book in which all measurements will be entered, a linen tape, an Abney clinometer and the rear end of the steel tape; he also collects the arrows inserted by the leader when use of each is completed. At the end of the line the total distance will be x tape lengths plus a residual length, e.g., x lengths of 20 m each plus a residual length of 17.62 m. The follower has x arrows, the leader $(10 - x)$ arrows, and the two totals must always be checked for error—the follower may have omitted to pull up one arrow during the operation.

Before moving the tape forward after the measurement of a length, the follower uses the Abney clinometer to take a slope measurement to the leader. If the two men are of equal height, it is sufficient to sight to the leader's eyes; if of different heights, to a point on the leader at the same height above the ground as the follower's eyes. If two Abneys are available, the leader may also carry one and may take reciprocal observations to the follower: comparison of the two readings provides a check and a mean value can be adopted. Both men should then walk along the tape from zero to 20 m (zero is always at the follower's end) to insert detail. The measurement

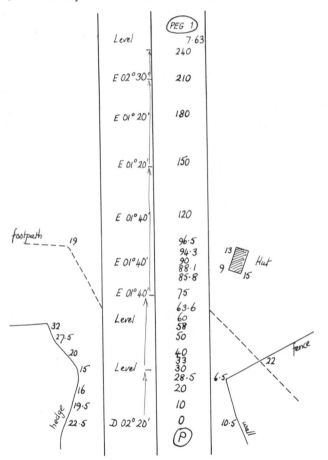

Fig. 32   Field book entries for chaining figures. Distances along the measured traverse line are entered upwards from the bottom of the page. Every chain length (in this case a 30-metre chain was used) is entered, with other distances required for plotting offset or other detail. Detail is entered to left or to right of this column as is appropriate. The second central column contains slope readings taken with the Abney clinometer over each chain length to allow for adjustment of slope distances to the horizontal. Even when slope angles are very small, and corrections for slope negligible, it is desirable to record Abney readings and individual chain lengths as a standard routine to ensure that no length is omitted in error.

at which the survey line crosses a stream, a footpath or a vegetation boundary, for example, must be recorded. Offsets are taken to detail on either side of the line, using the linen tape at right angles to the steel tape. In the measurement of offsets the leader takes the zero of the linen tape to the detail being mapped. The follower is responsible for seeing that the linen tape is at right angles to the traverse line, for reading the offset length and its distance along the steel tape, and for booking all the relevant information. It is only when both men are agreed that all detail has been recorded that a return is made to the ends of the steel tape and this is then moved forward in the normal way. Figure 32 illustrates the method of booking, in this case using a 30-metre tape. It is desirable to make an entry for every tape length even if no offsets to detail are taken, as the entry is a check on the total length of the line. In any case, a slope reading should be entered for every tape length, even if slope is uniform or negligible. It can be seen from the illustration that Abney readings are not always taken over tape lengths. If there is a clear break of slope, an Abney reading will be taken over each section. Figure 33 shows that a correction based on the

HORIZ. DIST.	CORRECTION			(B)			CORRECTION	HORIZ. DIST.
	2·102			360				
	3·153		26 30	340				
	0·005		E 26 30	310				
	0·030		02 30	305				
	0·030		02 30	275				
	0·030		02 30	245				
	0·030		02 30	215				
	0·055		E 02 30	185				
	0·165		06 00	175				
	0·165		06 00	145				
	0·165		06 00	115				
	3·570		D 06 00	85				
	4·284		31 00	60				
	4·284		31 00	30		A → B		
341·932	18·068		D 31° 00'	0		D 04°30'	1·116	358·884
				(A)				

Fig. 33 The centre column shows field book entries for measured distances, and slopes from *A* to *B*, and the left-hand column indicates correction for slope over each segment and the true horizontal distance *AB*. The right-hand column shows the correction applied to the overall measurement of 360 m for the single angle measured directly from *A* to *B*, with a resulting error of 16.95 m.

It is important always to correct for each slope segment.

In practice one would not use Abney readings to correct for slopes as steep as 26° 30′ or 31° 00′ as the instrument reads at best to only the nearest 5′ of slope. To obtain true horizontal distances correct to 0.01 m a more accurate determination of slope would be required.

overall slope angle from one end of the tape to the other does not necessarily equal the true correction calculated on readings over slope segments. There is no standard method of booking Abney readings, but the separate column to the left of the chainage distances is probably the best. If formlines, rather than contours, will be sufficient for the survey, Abney readings can also be taken over slopes to the side of the traverse line, formlines being calculated and plotted later (see p. 106).

Each leg of the traverse is measured in this way. At the start of each leg the leader takes over the arrows from the follower, and checks that the total is still 10. If the survey is to be completed with a stadi-altimeter, a telescopic alidade or some other form of range-finder, time can be saved by

Survey   *Theodolite traverse P–Q*
Date   *18 August 1973*
Observer *C.D. Jones*

AT	TO	FACE	HORIZONTAL CIRCLE ° ' "	HORIZONTAL ANGLE ° ' "	VERTICAL CIRCLE ° ' "	VERTICAL ANGLE ° ' "	TIME	LEVEL E O	REMARKS
P	M	L	176 04 10						
	Peg 1		221 15 40	45 11 30					
	M	R	356 03 50	45 10 50					
	Peg 1		41 14 40	45 11 10					
	P	L	163 27 00						
Peg 1	Peg 2		339 40 40	176 13 40					
	Peg 2	R	159 40 10	176 13 00					
	P		343 27 10	176 13 20					
	Peg 1	L	12 42 00						
Peg 2	Peg 3		238 36 50	225 54 50					
	Peg 1	R	192 41 50	225 55 10					
	Peg 3		58 37 00	225 55 00					

Fig. 34   The booking of horizontal angles on a theodolite traverse. Observations have been made with a 20″ instrument, giving readings by estimation to the nearest 10″.

Note that at each station the reading to the back station is subtracted from the forward reading to obtain the traverse angle, irrespective of the order in which they are observed or are entered in the field book. It is sometimes more convenient to take readings in the order shown for Peg 1 (if, for example, Peg 2 is difficult to see against a confusing background: it may more easily be found on Face Right while the telescope focus remains unaltered and while the observer can identify it for preliminary alignment with the open sights).

The sketch of the traverse line also helps by giving an indication of the approximate angle expected.

ignoring most offsets at the chaining stage and by recording only on-line detail.

Measurement of angles

At most of the traverse stations the theodolite is centred over the station mark and levelled, and the angle observed between the line to the back station and the line to the forward station. The angle is measured twice, once on Face Left and once on Face Right: the two measurements should agree within 60″ even with the use of a 20″ instrument, and should be closer with a more refined instrument. The mean value is taken. It should be noted that there is no need to observe the back station twice, as with a round of angles starting and closing on the R.O. during triangulation observation. The triangulation was the most accurate part of the survey, and the methods employed were the most rigorous. The theodolite traverse can afford to be a little less precise, as it is adjusted and held within the triangulation network. It is more precise than some of the methods that will be used at a later stage and which will be based on or between traverse points. Typical field book entries would be as in fig. 34. In calculating the angle, the reading to the back station is always subtracted from the reading to the forward station, irrespective of the order in which they are booked.

The first station in the traverse will be one of the triangulation points, and the angle measured will be between the back sight to another triangulation point and the forward sight to the first peg of the traverse. As the azimuth between the two triangulation points is known, the forward azimuth to the first peg can be calculated. The last station of the traverse will be another triangulation station, and an essentially similar procedure will be followed in reverse, observing the angle between the back line to the last traverse peg and the forward line to another triangulation station. If the list of azimuths over the triangulation network has been taken to the field as recommended (p. 36), it is possible to make in the field a calculation of azimuths over each line of the traverse and so to check for the misclosure between the values obtained for the final line as calculated from the traverse and from the triangulation. This misclosure should not exceed 2′ 30″, nor should the average correction to be applied at each station (obtained by dividing the total misclosure by the number of angles of which observations have been made) exceed 15″. If these values are exceeded, re-observation will be necessary. Error is most likely to be located at stations where the observed lines were short, as over short lines any displacement of the theodolite or the target pole, or any imprecision in the intersection of the target, will have a greater angular displacement. Over a short line observations should be made to a chaining arrow rather than to a pole, giving a more precise target for intersection, and centring of the theodolite is of particular importance. Very approximately, a centring displacement of 6 mm gives an angular difference of 1′ at a distance of 20 m. A second likely location of error is in any angle observed to a staff at a point above its base—non-verticality can obviously result in lateral displacement of the line of sight, with the introduction of angular error. Figure 35 illustrates the angular effect of observation to a non-vertical pole. Occasionally in rugged or wooded terrain it is impossible so to site traverse stations that the peg is always visible to the observer at the next station, but care must then be taken over verticality of the pole. If the assistant balances the pole lightly between his hands, an upright position is readily achieved. If only a small length is visible at the top of the pole,

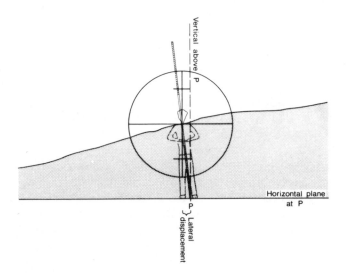

Fig. 35   It is important that target poles should be vertical. What is required is the angle to *P*, but if intervening obstructions conceal the lower part of the pole, as shown, and the pole is tilted, intersection of the lowest visible part of the pole will give an error of lateral displacement.
     Whenever possible, site traverse pegs so that the base of the pole is visible from adjacent stations.

it may be inadequate for the assessment of verticality, and if there is no means of avoiding the use of this station, it may be preferable to sight not to a pole but to an extended levelling staff held with its face to the observer, as the additional height of the staff provides a check on the vertical. If this has to be done, the staff should always face the observer, even though this presents a broader target for intersection than if the staff is held sideways: an extending staff tends to have its upper section leaning forward when the base is vertical, which gives a displacement error when viewed sideways but no error when viewed from the front.

     If the traverse closes back to its starting point, the initial angular observations must include not only another triangulation point but also both first and last traverse pegs. This allows calculation of the initial bearing and of the final misclosure. Figure 36 shows a field book entry for such a station. In calculating angles, the reading to the back station is always subtracted from

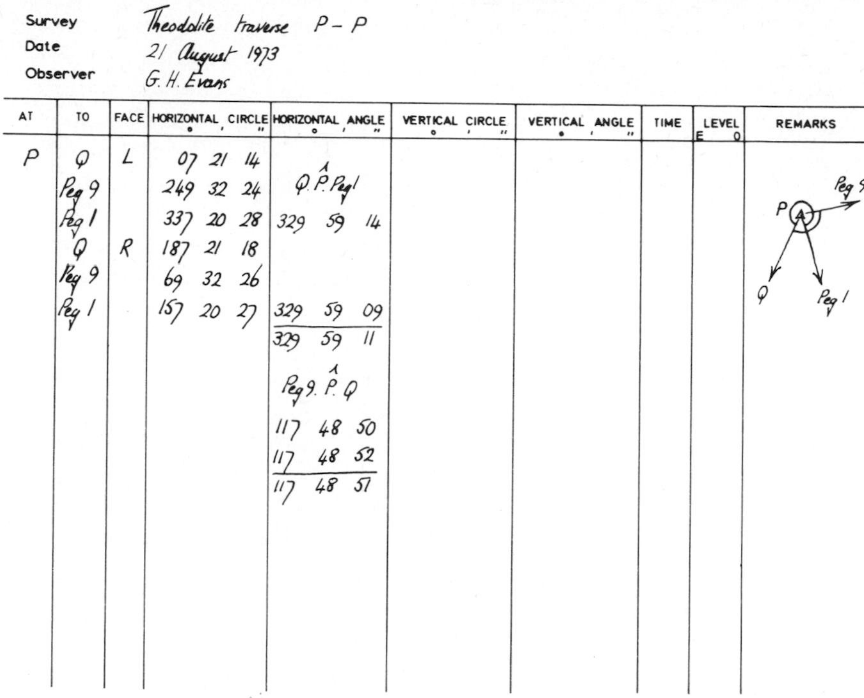

Survey   Theodolite traverse P–P
Date   21 August 1973
Observer G. H. Evans

AT	TO	FACE	HORIZONTAL CIRCLE ° ′ ″	HORIZONTAL ANGLE ° ′ ″	VERTICAL CIRCLE ° ′ ″	VERTICAL ANGLE ° ′ ″	TIME	LEVEL E O	REMARKS
P	Q	L	07 21 14						
	Peg 9		249 32 24	Q.P̂.Peg1					
	Peg 1		337 20 28	329 59 14					
	Q	R	187 21 18						
	Peg 9		69 32 26						
	Peg 1		157 20 27	329 59 09					
				329 59 11					
				Peg9.P̂.Q					
				117 48 50					
				117 48 52					
				117 48 51					

Fig. 36 Where a traverse closes back to its starting point, observe all necessary angles together as shown. As before (see fig. 34) the sketch indicates the angles being measured, in this case taking the initial bearing from the known line *PQ* and closing back to the same line as a check on the angular observations round the traverse.

In this example angles have been measured with a 1″ theodolite.

the reading to the forward station, so that in this example *Q* is subtracted from Peg 1, and Peg 9 is subtracted from *Q*. The angular misclosure is then obtained in the usual way. Note that the final angle is from Peg 9 back to *Q* and not to Peg 1. The latter would permit calculation of the misclosure round the closed traverse but would not provide any check on, or adjustment of, the angle $Q \cdot \hat{P}$. Peg 1. As *PQ* will be a relatively long line it is the least subject to intersection or displacement errors and is therefore most suitable as the line over which the misclosure is calculated.

Calculation of traverse

Calculation is relatively simple, requiring only the ability to use logarithms and to do simple addition and subtraction. Figure 37 illustrates a computation form for a short traverse with six observed angles. The initial azimuth is taken from the calculated triangulation network, in which the azimuth *PM* has been determined as 127° 43′ 07″. As the line to *M* is the back sight at *P*, this is entered in the second column as the back bearing. The angle at *P* between the lines to *M* and to Peg 1 was 45° 11′ 10″. This added to the

Survey *Traverse P–Q*
Date *18 August 1973*
Computer *C. D. Jones*

Final Coordinates (Station *Q*) E 5785·80   N 2793·81
Initial Coordinates (Station *P*) E 6241·70   N 3872·00

FROM STATION	BACK BEARING / OBSERVED ANGLE / FORWARD BEARING	CORRN TO BEARING BRG.	CORRECTED BEARING	TRUE HORIZ. LENGTH	TOTAL DISTANCE TRAV/SD	LOG. SIN BEARNG / LOG. LENGTH / LOG ΔE	LOG COS BEARNG / LOG. LENGTH / LOG. ΔN	E	W	N	S	EASTINGS CORRECTION CORR. EASTINGS	NORTHINGS CORRECTION CORR. NORTHINGS	TO STATION
P	127° 43' 07" (Peg M) / 45 11 10 / 172 54 17	+7"	172 54 24	247·53	248	9.094 6678 / 2.393 6278 / 1.488 2456	9.996 6433 / 2.393 6278 / 2.390 2911	30·57			245·64	6272·27 / +0·06 / 6272·33	3426·36 / −0·03 / 3426·33	Peg 1
Peg 1	352 54 17 / 176 13 20 / 169 07 37	+15"	169 07 52	393·42	644	9.275 4547 / 2.594 8544 / 1.870 3111	9.992 1386 / 2.594 8544 / 2.586 9950	74·18			386·36	6346·45 / +0·16 / 6346·61	3240·00 / −0·09 / 3239·91	Peg 2
Peg 2	349 07 37 / 225 55 00 / 215 02 37	+23"	215 03 00	202·47	843	9.759 1321 / 2.306 3607 / 2.065 4928	9.913 0989 / 2.306 3607 / 2.219 4596		116·28		165·75	6230·17 / +0·21 / 6230·38	3074·25 / −0·12 / 3074·13	Peg 3
Peg 3	35 02 37 / 210 38 40 / 245 41 17	+3"	245 41 48	347·03	1190	9.959 6992 / 2.540 3670 / 2.500 0662	9.614 4409 / 2.540 3670 / 2.154 8079		316·28		142·83	5913·89 / +0·30 / 5914·19	2931·42 / −0·17 / 2931·25	Peg 4
Peg 4	65 41 17 / 157 22 00 / 223 03 17	+38"	223 03 55	188·08	1378	9.834 3133 / 2.274 3426 / 2.108 6559	9.863 6455 / 2.274 3426 / 2.138 0081		128·43		137·41	5785·46 / +0·34 / 5785·80	2794·01 / −0·20 / 2793·81	Q
Q	43 03 17 / 282 42 10 / 325 45 27	+46"	325 46 13 (LJ)					104·75 / 560·99 / 104·75 / 456·24		−	1077·99			

Misclosures   ΔE 0·34   ΔN 0·20

$$\text{Accuracy} = \frac{\sqrt{(0·34)^2 + (0·20)^2}}{1378} = \frac{0·393}{1378} = \frac{1}{3506}$$

Fig. 37   Calculation of a theodolite traverse (see text). Note: Corrections are usually made to the nearest 0.1 m, which is adequate for plotting purposes. The calculation here has been made to the nearest 0.01 m in order to make clear the approximations given in the text.

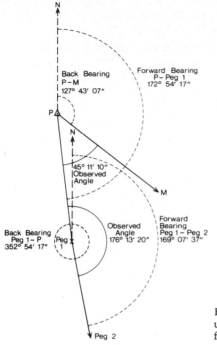

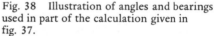

Fig. 38 Illustration of angles and bearings used in part of the calculation given in fig. 37.

back bearing gives the forward bearing to Peg 1 as 172° 54′ 17″ (fig. 38). If the forward bearing from *P* to Peg 1 is 172° 54′ 17″, the bearing in the opposite direction from Peg 1 to *P* will differ from this by 180°, i.e. it will be 352° 54′ 17″, and this is entered on the next line as the back bearing from Pcg 1. The observed angle at Peg 1 was 176° 13′ 20″: this added to the back bearing gives the forward bearing to Peg 2 of 169° 07′ 37″. At Peg 2 the back bearing will differ from this by 180°, i.e. it will be 349° 07′ 37″. The calculation continues to the final station occupied, *Q*, with a calculated bearing from *Q* to *J* of 325° 45′ 27″. However, the bearing *QJ* is known from the network observations and calculations, and is 325° 46′ 13″. The traverse misclosure is therefore 46″, and a correction of 46″ needs to be applied to the final bearing to adjust it to the known figure. It is assumed that error is spread evenly throughout the traverse, so that 46″ has to be spread over six angles. This is done (in this case) by adding 7″ to each of the two smaller angles , and 8″ to each of the four larger angles. This could be done by entering in column 4 revised angles of 45° 11′ 17″, 176° 13′ 28″, 225° 55′ 08″, etc., and repeating the arithmetical calculation as before. However, it is easier to apply to each forward bearing an appropriate correction, as shown. The forward bearing from *P* to Peg 1 will obviously be greater by 7″, as the angle at *P* has been increased by 7″. The forward bearing at Peg 1 will be greater by 15″ (representing the 7″ on the angle at *P* plus the 8″ on the angle at Peg 1), the forward bearing at

Peg 2 will be greater by 23″, and so on. The total misclosure and the correction per station are both within the permitted limits.

Once adjustment of bearings is completed, lengths can be considered. Correction of measured lengths to the horizontal can be done in the field book or the traverse computation form. In the example shown, the line from *P* to Peg 1 consisted of eight tape lengths each of 30 m, plus a residual distance of 7.63 m (see fig. 32). The vertical angles for each section are entered, and the appropriate slope correction taken from Table 3. The corrections are always negative, being reduction of the slope distance to the horizontal distance. In the table below, *D* or *E* signifies Depression or Elevation: the letter makes no difference to the correction, but may be used later for formline computation if accurate contours are not required. At this stage, ignore the column headed "Total Distance Traversed".

	ΔP − Peg 1		
Measured distance	*Slope*	*Correction*	*Horizontal distance*
7.63	—	—	7.63
30.00	E.02° 30′	.03	29.97
30.00	E.01° 20′	.01	29.99
30.00	E.01° 20′	.01	29.99
30.00	E.01° 40′ ⎫		
30.00	E.01° 40′ ⎬	.03	74.97
15.00	E.01° 40′ ⎭		
15.00	—	—	15.00
30.00	—	—	30.00
30.00	D.02° 20′	.02	29.98
247.63		.10	247.53

For each leg of the traverse, the log sine bearing and log cosine bearing are then extracted from the tables and inserted in the centre columns, with the logarithm of the horizontal length. The sums of the logarithms give the difference in eastings and northings for each leg, as for example between *P* and Peg 1 (see fig. 37). These differences are entered in the appropriate columns, *E* or *W* and *N* or *S* respectively. The co-ordinates of each traverse peg can then be calculated: the co-ordinates of *P* are known, the difference in co-ordinates between *P* and Peg 1 are calculated, the co-ordinates of Peg 1 are found, and so on throughout the traverse. Finally the co-ordinates of *Q* are calculated, and compared with the known co-ordinates from the triangulation network. The misclosure is seen to be −0.34 in eastings and +0.20 in northings. The closing error equals $\dfrac{\sqrt{\Delta E^2 + \Delta N^2}}{\text{Total Distance Traversed}}$ where ΔE and ΔN are the misclosures in eastings and northings respectively, and in this class of work the error should not exceed 1: 3000. To check the misclosure we require the total distance, and at this point the column headed "Total Distance Traversed" can be completed by adding each successive horizontal length, leg by leg through the traverse. It is best not to complete this column until this stage, in case the running total should inadvertently be mistaken for the length of a leg in the extraction of logarithms. If the error is within the acceptable limits, it is assumed that it has accumulated evenly throughout the traverse, and adjustment will be made accordingly. This is known as Bowditch's Rule, which is also used in the adjustment of compass traverses, and although it makes assumptions which cannot be justified in full, it is adequate for this class of work. The Bowditch Rule states that at any traverse station

Correction in Eastings = Total Misclosure in Eastings

$$\times \frac{\text{Total distance traversed to the station}}{\text{Total length of traverse}}$$

For Northings a similar correction can be applied, substituting Northings for Eastings.

The total length has been calculated to obtain the accuracy of the traverse. In the course of the addition the total distance traversed to each peg is also calculated (as a check, the sum of the horizontal lengths should equal the final figure of distance traversed). The Eastings misclosure is 0.34 m in a total distance of 1378 m, which means that there should be an adjustment of 0.01 metres for every 40 m traversed. At Peg 1 the distance traversed is 248 m, and so the correction is +0.06; at Peg 2 the distance traversed is 641 m, and the correction is +0.16 m; at Peg 3 the distance traversed is 843 m, so that the correction is +0.21 m; and so onwards through the traverse. The Northings misclosure is 0.20 m in a total distance of 1378 m, requiring an adjustment of 0.01 m for every 69 m traversed. At Peg 1 the correction is therefore −0.03, at Peg 2 it is −0.09, at Peg 3 it is −0.12, at Peg 4 it is −0.17, and at Q the full correction of −0.20 must obviously be applied. Note that in this example calculation has been made to the nearest 0.01 m, which is more accurate than is usually required, but it demonstrates the full adjustment. It is usually sufficient to calculate to the nearest 0.1 m.

If the misclosure is too great (i.e. if accuracy does not reach 1 : 3000) one should first look for gross error. There are two likely probabilities (excluding misreading of the tables):

(a) transposition of the log cos and the log sin;
(b) an error of one tape length not being recorded.

(a) A check can be made by inspection of the computation. The line from P to Peg 1 is on a bearing of approximately 173°, so that the difference in co-ordinates should include an S movement (Southward) not very different from the length of the leg, and a small E movement. Much the same applies

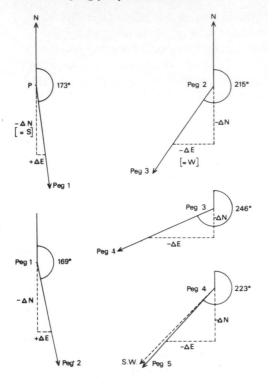

Fig. 39   The bearings of the traverse legs recorded in fig. 37. The legs are not drawn to scale. The relative proportions of ∆E and ∆N over each leg can be seen as a check on the calculations.

to the next leg on 169°. The third leg on 215° is between South and South-West: therefore the S figure should be greater than the W figure. The fourth leg on a bearing of nearly 246° should have a W figure considerably larger than the S figure; the final leg from Peg 4 to Q is on 223°, very close to the 225° of South-West, but with the S figure slightly greater than the W figure. Figure 39 illustrates the ∆E and ∆N proportions for each of these bearings. Mistakes are usually detected easily at this stage. If the error is not due to transposition of the sine and cosine, it is frequently due to arithmetic, often with a summed logarithm written as 2.xxx xxxx instead of as 1.xxx xxxx, for example. Table 6 gives a very basic form of Traverse Table which shows differences in Eastings and Northings for a length of 100 m on different bearings, and this can be used for a quick check, although not as a substitute for full calculation.

(b) If checking of the whole computation fails to reveal any arithmetical errors, it is probable (if the error is large) that the taped length on one leg was inaccurately recorded. It is frequently possible to identify the leg on which the error occurred. If in the example illustrated the calculated co-ordinates of Q had been 5812.90 Eastings and 2806.20 Northings, the

misclosure would have been +27.10E and +12.40N. An error of this magnitude is less disturbing than an error of 1 or 2 m, as the latter usually
represents poor work in the field and may necessitate repetition of the
complete survey. The misclosure figures can be plotted to give a triangle
as in fig. 40, and it is seen that the bearing of the hypoteneuse is roughly

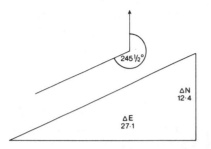

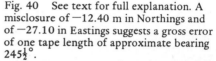

Fig. 40   See text for full explanation. A
misclosure of −12.40 m in Northings and
of −27.10 in Eastings suggests a gross error
of one tape length of approximate bearing
$245\frac{1}{2}°$.

midway between south-west and west, i.e. midway between 225° and 270°.
On this bearing the most likely leg for error is that on nearly 246° from
Peg 3 to Peg 4. As a 30-metre steel tape was used, the likelihood is a gross
error of 30 m. The traverse table (Table 6) shows that 30 m on a bearing
of 246° would give approximately −27.4 in Eastings and −12.2 in Northings.
An extra 30-metre length on this bearing would thus reduce the misclosure
to −0.3 in Eastings and +0.2 in Northings, which would be acceptable. A
check against the paces recorded on the reconnaissance of the traverse
should confirm the discrepancy, and the necessary field check over this
line should reveal the error immediately. Had the misclosure in this example
been +4.0 in Eastings and −29.7 in Northings, it would have indicated a 30-
metre error a little to the east of south, i.e. about 170°. In this case the first
two lines would be suspect, and P to Peg 1 and Peg 1 to Peg 2 would both
need checking against the reconnaissance and in the field. If the reconnaissance figures suggest identification of the error in, for example, the line from
Peg 1 to Peg 2, it may save much time if calculation and plotting proceed on
the assumption that the location of the error is known and the traverse is
amended accordingly. This may allow the observer to return to the field on
the following morning with the traverse plotted on a plane table and ready
for the start of detail mapping, provided that his overnight work is confirmed with a check measurement at the start of the day.

Another precaution against gross error is the use of stadia observations
in the course of angular observation. After completed measurement of the
horizontal angles, a levelling staff can be substituted for the target pole
before vertical angles are measured. The observer uses a tape to measure the
height of the trunnion axis above the ground; he points the telescope at the
target staff and brings to centre the sensitive bubble on the theodolite; he
uses the vertical tangent screw to bring the crosswire to the staff reading
equal to the trunnion axis height above the ground; he reads and records
upper and lower stadia; and he records the vertical circle reading. The stadia

readings can be used with table 4A to give a close approximate horizontal distance to the target position, and this distance can be used as a check against gross error in the measured length.

The traverse pegs can be plotted on squared paper from their calculated co-ordinates, on the scale required, and will then usually be transferred to a plane table sheet for subsequent work. The transfer is best effected with a tracing table, if available. If working under field conditions, the points can be transferred to tracing paper and thence to the plane table sheet. The final move should never be by "pricking through" with a pin, which mutilates both map and plane table: shading of the underside of the paper, with a soft pencil, at each of the traverse points, will enable the tracing paper to be used directly, as cross strokes on the point positions will then be marked on the plane table sheet beneath. The squared paper on which the traverse is plotted should already have marked on it the network of triangulation points which will also be transferred as required.

Traverse heights

If possible, traverse pegs should be included in the levelling programme described on page 44, so that the heights are known accurately. If, however, accurate contours are not essential but reasonably accurate formlines will be required, use can be made of vertical angles during the traverse. As with triangulation observations, each forward and backward station will be observed for a vertical angle on Face Left and Face Right. The mark for intersection will in each case be a mark on the target pole or staff at the same height above the ground as is the trunnion axis of the theodolite. This is best achieved by observing to a levelling staff as suggested above, taking stadia readings also. The vertical angular reading required is, of course, that obtained with the telescope crosswire on the staff at trunnion axis height. Face Left and Face Right readings are taken, and they should agree within about 1 minute of arc, e.g.

FL Vertical circle reading $356°\ 27'\ 10''$ = angle of $03°\ 32'\ 50''$
FR Vertical circle reading $183°\ 33'\ 40''$ = angle of $\underline{03°\ 33'\ 40''}$
Mean $\ \ 03°\ 33'\ 15''$

Corresponding readings over the line in the opposite direction could be

FL Vertical circle reading $\ 03°\ 31'\ 50''$ = angle of $03°\ 31'\ 50''$
FR Vertical circle reading $176°\ 27'\ 20''$ = angle of $\underline{03°\ 32'\ 40''}$
Mean $\ \ 03°\ 32'\ 15''$

Before intersection with the crosswire the sensitive bubble is always brought to centre (or the bubble ends are read against graduations of known values of dislevelment, see p. 9). The mean of the vertical angles from opposite ends of the line can be accepted if they agree within reasonable limits (the limits vary with length of the line, being greater over short lines). For lines 200 m long, a difference in vertical angle of $10'$ makes a difference in calculated altitude of approximately half a metre. A 10-minute difference, when meaned, will affect the calculated altitudes in the traverse by 0.25 m; a 5-minute difference will correspond to 0.125 m, etc. The accuracy required will therefore determine the degree of correspondence considered acceptable, with wider variation possible over short lines than over long lines. Use of table 4B will enable the surveyor to work out the acceptable limit over any length of line, in relation to his desired accuracy.

Over each leg of the traverse the difference in height can be calculated as

Difference in height = Horizontal distance x Tangent of vertical angle

care being taken that elevations and depressions are correctly distinguished. The true horizontal distances can be obtained from the traverse computation after correction for slope. In this way the altitudes of all the traverse pegs can be computed. There will be a discrepancy between calculated and known height at the closing point, e.g. station *Q* in the example. Bowditch's rule can again be applied for adjustment of heights in this class of work, correction for height error being proportional to the distance traversed. In the example used in fig. 37, a misclosure of +0.018 m at *Q* would be adjusted as 0.0013 per 100 m, giving correction values of 0.003, 0.008, 0.011 and 0.015 to be subtracted at Pegs 1–4 respectively.

If the misclosure is unduly large, a check can be made by calculating the total difference in height along each leg of the traverse, using either

Difference in height = Horizontal distance x Tangent of vertical angle

or

Difference in height = Slope distance x Sine of vertical angle

and calculating separately each chain length with its corresponding Abney level reading. If elevations and depressions have been entered against these readings, this procedure, although laborious, provides a fairly reliable check on the vertical angles. It also provides a series of spot heights along the traverse line, and these may be useful in mapping detail at a later stage.

By far the simplest and quickest calculation, or check, is made by reading stadia lines as described on page 68, and extracting height differences from table 4B.

If only generalised formlines are required, these can be added to the map from Abney clinometer observations made during a theodolite traverse, but the method is better suited to the standard of accuracy of a compass traverse, and is therefore described on page 103.

Mapping of detail with the theodolite

A theodolite of standard design measures horizontal and vertical angles. The taking of stadia readings to a graduated staff, as already described, permits its additional use as a tacheometer, for the optical measurement of distance. A reference was made on p. 46 to the use of stadia readings as a check during levelling operations, and further reference has been made

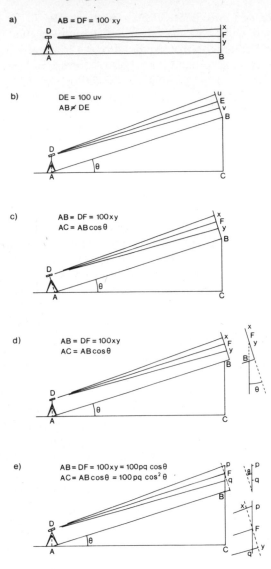

Fig. 41   Optical measurement of distance by use of stadia readings.
(a) With a level telescope and horizontal line of sight.
(b) With an inclined telescope vertically above *A* the staff would need to be held at an angle of 90° to *AB*, giving stadia readings *u* and *v*. But *DE* no longer equals *AB*.
(c) Unfortunately it is not possible to set the theodolite also at right angles to *AB*.
(d) It would be very difficult for the staff holder to keep the staff at 90° to the line of sight while maintaining *F* vertically above *B*.
(e) In practice the theodolite *D* must be vertically above *A* and the staff must be vertical at *B*. Then

$$AC = 100 \cdot pq \cdot \cos^2\theta$$
$$BC = 100 \cdot pq \cdot \frac{\sin 2\theta}{2}$$

above to the use of stadia readings to check horizontal distances and differences in height. This use of stadia readings for the optical measurement of distances is known as tacheometry. The spacing of the stadia lines is designed to give a 1:100 ratio for the stadia intercept on the staff, held vertically, and the horizontal line of sight from the telescope to the staff. This 1:100 ratio holds only for readings taken with the telescope horizontal and the staff vertical. If the line of sight is inclined, the ratio holds true only if the staff is tilted to remain at right angles to the line of sight as in fig. 41c, and the distance obtained is no longer the horizontal distance from telescope to staff, but is the distance on the slope. To obtain the true horizontal distance this value must be corrected for the angle of slope. The horizontal distance AC in fig. 41c $= AB \cos \theta$, where θ is the angle of slope, but this also requires the theodolite to be set off the vertical. In any case it is not easy for the staff holder to tilt the staff by exactly the right amount, judging by eye: it is much easier for him to maintain verticality of the staff as in fig. 41e. In this case

(a) the interval from the lower stadia reading q to the horizontal crosswire no longer equals the interval from the crosswire to the upper stadia reading p, as the point q is nearer to the telescope.

(b) the interval pq between lower and upper stadia will be greater than if the staff were at right angles to the line of sight, i.e. xy, and the slant range distance will appear to be too great. The *slant range* is the name given to the value $100 . pq$.

It can be seen from fig. 41e that $pq = xy . \sec \theta$

whence $$xy = \frac{pq}{\sec \theta} = pq . \cos \theta$$

The slope distance $\quad AB = 100 . xy = 100 . pq \cos \theta$

and the true horizontal distance $AC = AB \cos \theta$

$$= 100 . pq \cos^2 \theta$$

$$= 100 . pq(1 - \sin^2 \theta)$$

To obtain the true horizontal distance, therefore, the slant range has to be reduced by a correction of $100 . pq . \sin^2 \theta$. Table 4A tabulates values of $100 . pq . (1 - \sin^2 \theta)$ for angles from $0°$ to $45°$ and for values of pq from 1 to 9. If the horizontal distance is required only to the nearest quarter of a metre, the values obtained from the table will not introduce any plottable error.

It may also be desirable to know the difference in altitude between A and B, i.e. the vertical interval BC.

$$BC = AB \sin \theta = 100 . xy . \sin \theta$$

$$= 100 . pq . \cos \theta \sin \theta$$

$$= 100 . pq . \frac{\sin 2\theta}{2}$$

Table 4B lists values of $100 \cdot pq \cdot \dfrac{\sin 2\theta}{2}$, also for angles from $0°$ to $45°$, in a similar form for easy extraction of height differences. Both tables reduce the likelihood of error in the field by substituting addition for multiplication. Most observers will find it difficult with a standard theodolite to read staff divisions above 200 m, but the tables are set out in columns up to 900 units to allow both easy extraction of figures and also use up to 900 ft if metric units are not being employed.

These optical techniques of length and height determination can be applied in a number of ways to theodolite traverses.

Addition of detail to standard traverses

Traverse legs are measured as before. At the first station the theodolite is used in the normal manner to measure the necessary horizontal and vertical angles. When these standard observations have been completed, detail within 200 m of the theodolite can be observed for direction, distance and altitude. The uncorrected back bearing at the station is known, for example, $349° \ 07' \ 37''$ over the line Peg 2–Peg 1 in fig. 37. With the horizontal circle clamped, the upper plate of the theodolite is turned until the micrometer for horizontal circle readings records $349° \ 07' \ 37''$ (or, if a 20'' instrument is being used, until it records approximately $349° \ 07' \ 40''$). The upper plate is then clamped, and the lower plate released: the telescope may now be pointed in any direction without altering the micrometer reading. The point Peg 1 is intersected using the *lower* plate clamp and tangent screws, so that the reading remains unaltered. A check is made to ensure that with Peg 1 intersected the micrometer reading is still $349° \ 07' \ 40''$. Henceforth the lower plate remains untouched, all subsequent observations and intersections being made with the use of the upper plate controls only. It should be clear that as the circle gives the correct azimuth value when the telescope is directed at Peg 1, the circle reading when any other point is intersected will be the true azimuth from the observer to that point.

A short length of Scotch tape is placed across a levelling staff at a height equal to that of the trunnion axis of the theodolite above the ground (the staff is held vertically alongside the instrument for this purpose). In all observations for stadia readings p and q, the horizontal crosswire is brought onto this tape, a precaution necessary to avoid corrections for the height above ground to the theodolite telescope or the intersected points on the staff. The assistant moves to any point of detail required, for example the Monument north of Peg 2 (see figs. 43 and 47), and holds the staff vertical. The observer brings the staff into the field of view of the telescope, and completes the intersection of the tape using the upper plate and vertical circle tangent screws. He then reads the upper and lower stadia, checks that the intervals of the stadia from the tape are roughly equal (they will, of course, be equal only when the telescope is horizontal), and signals to the assistant to move to the next point. While the assistant moves, and before altering the instrument controls, the observer reads the vertical circle by estimation to the nearest 10 minutes of arc. How he does this depends on the make and model of the theodolite. If using a micrometer instrument, for example, he would previously have set the micrometer scale at zero, and he would make his estimate of the vertical angle by noting the position of the index mark on the main scale (see fig. 42). Having read and booked the vertical angle, the observer looks at the appropriate line in table 4A and reads the value of $100 (1 - \sin^2 \theta)$ for the slant range, e.g., if the stadia readings are 2.420 and

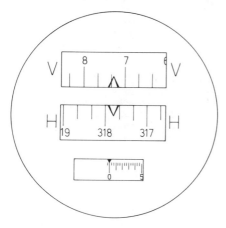

Fig. 42 The micrometer is set to read zero minutes and zero seconds in the bottom window. It is then easy to estimate vertical and horizontal circle readings to the nearest 10 minutes. In the example shown, the readings would be $07° 20'$ and $317° 50'$.

1.160 m, the slant range is 126.0 m; the vertical angle is read as $+07° 20'$; from table 4A the true horizontal distance will be $98.37 + 19.67 + 5.90$, and the horizontal distance is thus 123.94 m. The observer then reads the horizontal circle to the nearest 10 minutes of arc (by estimation with the micrometer scale at zero as before) and enters this figure. Then, or later, he also looks at table 4B for $07° 20'$ and takes out the value for 126 m, namely $12.66 + 2.53 + 0.76 = 15.95$. This is the difference in altitude between the instrument station Peg 2 and the staff position. As Peg 2 is known to be at an altitude of, say, 132.16 m, the altitude of the staff position is 148.11, as the angle $07° 20'$ was one of elevation. By the time the observer has completed these observations, the assistant is ready at the next point to be fixed. The observer turns the telescope to him, intersects the tape on the staff, reads the upper and lower stadia, checks the intervals, and signals for the next move. He then continues his readings and calculations as before. Note that the Tables are used to two decimal places even though the slant range can usually be read only to the nearest half-metre over long lines: this is necessary because the final figure used is the sum of three others taken from the table and error due to the rounding-off of each figure to one place of decimals could be cumulative. It is unrealistic to use a final horizontal distance expressed in greater detail than the nearest tenth of a metre.

At the end of the operation the observer has booked perhaps several dozen points around his traverse station. For each there is a true bearing, a

Survey   Theodolite traverse P-Q
Date   18 August 1973
Observed   C.D. Jones

AT	TO	FACE	HORIZONTAL CIRCLE ° ' "	UPPER STADIA	LOWER STADIA	I	SLANT X 100 = RANGE	VERTICAL ANGLE ° ' "	HORIZ. DISTAN.	HEIGHT DIFF.	REMARKS
Peg 2	Peg 1	L	369 07 40								Back bearing
	Peg 3		215 02 40	2.520	0.490	2.030	203	+ 03 10	202.4	+11.16	Forward bearing
	2A		10 40	2.110	0.910	1.200	120	− 05 10	119.0	−10.68	Edge of bare rock outcrop
	2B		21 30	2.020	1.010	1.010	101	− 04 10	100.5	− 7.32	"
	2C		27 50	1.750	1.275	0.475	47.5	− 02 30	47.4	− 2.07	"
	2D		46 00	1.870	1.145	0.725	72.5	− 01 00	72.5	− 1.26	"
	2E		57 10	1.950	1.070	0.880	88	Level	88	—	"
Tape at 1.510	2F		74 10	1.990	1.035	0.955	95.5	+ 00 50	95.5	+1.39	"
	2G		83 30	2.310	0.705	1.605	160.5	+ 01 00	160.5	+2.88	"
	2H		89 10	2.300	0.705	1.595	159.5	+ 01 30	159.4	+ 4.17	Footpath
	2J		92 50	2.060	0.960	1.100	110	+ 01 30	109.9	+2.87	"
	2K		114 40	1.965	1.060	0.905	90.5	+02 00	90.4	+3.15	"
	2L		151 00	1.950	1.065	0.885	88.5	+00 30	88.5	+0.77	"
	2M		194 30	2.210	0.810	1.400	140.0	− 00 50	140.0	− 2.03	"
	2N		246 00	2.460	0.560	1.900	190	− 04 10	189.0	−13.70	Stream
	2O		252 20	2.110	0.905	1.205	120.5	−02 20	120.3	− 4.87	"
	2P		272 20	1.940	1.080	0.860	86	− 01 50	85.9	− 3.39	Spring
	2Q		280 10	2.275	0.735	1.540	154	− 01 10	153.9	− 3.14	Wall
	2R		287 40	2.030	0.990	1.040	104	Level	104	—	Corner Wall
	2S		329 50	2.430	0.590	1.840	184	−03 50	183.1	−12.22	Wall
	2T		357 40	1.810	1.210	0.600	60	+00 10	60.0	+ 0.18	Post
	2U		359 00	2.140	0.880	1.260	126	+07 20	123.9	+15.95	Monument

Fig. 43   Field book entries for stadia observations of detail points around Peg 2 on a theodolite traverse. The points are shown in fig. 47.

calculated horizontal distance and an altitude. When he has calculated and plotted his traverse, he can plot on the map each point of detail. The mapping of detail can thus continue even in rain, if a surveying umbrella is stuck in the ground to shelter the theodolite. A sketch of the detail points should be made by the observer and by the assistant, with each point numbered or lettered on a pre-arranged code. The two sketches check each other for identification of plotted points.

With an effective range of about 200 m, mapping of detail round each traverse point in this way can complete a wide belt along each traverse line, and the details are preserved for remapping on other scales if required, provided that the scale chosen does not have a plottable error smaller than 0.25 m.

If the Scotch tape fixed on the levelling staff at telescope height cannot be seen through the telescope because of obstructions, a second piece of tape should be attached to the staff at some distance, e.g. 2.000 metres, above the first tape. Intersection of the higher tape, and stadia readings above and below it together with the new vertical angle, will allow calculation of the horizontal distance to the staff, and a figure for the altitude of the staff base which will be too high, in this case by 2.000 m. Employment of this upper tape is extremely useful when vegetation is high. The observer

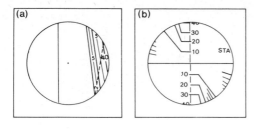

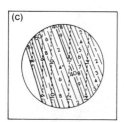

Fig. 44   The Ewing stadi-altimeter: eyepiece images.

(a) Before adjustment. The light dot (shown here by a black spot) lies to one side of the vertical axis of the graph. A red line on the graph is shown here by a broken line numbered 2.

(b) Adjustment completed. The light dot has been brought on to the vertical axis and the drum rotated until the dot lies also on the horizontal axis. Note that the black graph lines are cut off short to avoid congestion near the origin, so that readings cannot be taken when the telescope is close to the horizontal. However, distances can still be obtained from stadia readings, as correction is close to zero (no red lines are visible). See the text for the method of getting altitude differences in this situation.

(c) Black lines on the graph are broken only for figures. Red lines, which on the graph are also continuous between figures, are here shown by broken black lines. The drum has been rotated so that the light dot records 42.2 on the black graph, when the reading from the red lines is 11.3, which is the correction to be applied to the slant range to bring this down to the true horizontal distance.

must be careful to note which observations are made to the upper tape, thereby needing correction of the altitude value obtained.

Direct-reading tacheometric instruments

The method described above for the optical measurement of distances and of differences in altitude can be simplified by the use of instruments which are specially designed to give direct readings of at least some of the data required.

One such instrument is the Watts Microptic Theodolite No. 1 with a Ewing Stadi-altimeter attachment (fig. 46). This theodolite is a standard instrument with a micrometer reading. It can be modified by the addition of a bracket in place of the open sights on top of the telescope in the Face Left position, and the fitting when required of the stadi-altimeter. The latter consists of two parts, one being a small eyepiece attachment which fits on the bracket, the other being a rectangular metal box which fits on the top of the right-hand standard. The box houses a cylindrical drum on which is mounted a

Survey   Theodolite traverse with Stadi-altimeter P–Q
Date   18 August 1973
Observer   C.D. Jones

Peg 3   279.2 metres
Peg 4   250.1 metres

AT	TO	FACE	HORIZONTAL CIRCLE ° ' "	UPPER STADIA	LOWER STADIA	I	SLANT X100 = RANGE	SLOPE CORR.	HORIZ. DISTANCE	HEIGHT SCALE	SPOT HEIGHT	REMARKS
Peg 3	Peg 2	L	35 02 40									Footpath
	3A		160 00	2.370	0.650	1.720	172	0.4	171.6	92.2	271.4	Footpath
	3B		197 40	2.160	0.930	1.170	117	0.7	116.3	91.3	270.5	"
	3C		220 00	2.150	0.870	1.280	128	2.8	125.2	81.3	260.5	" junction
Tape at	3D		243 30	2.255	0.765	1.490	149	3.7	145.3	77.0	256.2	"
1.510	3E		210 20	1.705	1.320	0.385	38.5	0.9	37.6	93.9	273.1	"
	3F		250 30	2.335	0.685	1.650	165	3.6	161.4	75.8	255.0	Path/stream crossing
	3G		257 40	2.310	0.710	1.600	160	3.5	156.5	76.6	255.8	Stream
	3H		264 30	2.335	0.685	1.650	165	3.0	162.0	77.9	257.1	"
	3K		277 10	2.265	0.760	1.505	150.5	3.2	147.3	78.2	257.4	"
	3L		323 30	2.775	0.245	2.530	253	0.7	252.3	86.8	264.0	Wall
	3M		312 00	3.900	0.100	3.800	380	0.2	379.8	92.7	271.9	" . Road 0.5m above tape
	4A		289 20	unreadable at long range								
	Peg 4		245 41 40									
Peg 4	Peg 3	L	65 41 40									
	4A		334 40	3.755	0.435	3.320	332	1.1	330.9	19.0	269.1	Road. Reading 0.5 high
	4B		334 00	2.705	0.485	2.220	222	1.0	221.0	14.7	264.8	Spring
	4C		335 00	2.370	0.820	1.550	155	0.7	154.3	10.5	260.6	Stream
	4D		340 10	2.145	1.035	1.110	110	0.6	109.4	7.9	258.0	Stream/path crossing
	4E		312 20	2.370	0.825	1.545	154.5	1.4	153.1	14.7	264.8	Path/road junction
Tape at	4F		317 50	2.430	0.765	1.665	166.5	1.5	165.1	15.1	265.2	Road bend
1.595	4G		340 00	1.880	1.315	0.565	56.5	0.1	56.4	1.9	252.0	Stream
	4H		225 00	1.705	1.485	0.220	22	0.1	21.9	98.9	249.0	"
	4J		175 30	2.795	2.400	0.395	39.5	0.1	39.4	98.4	248.5	" Reading 1.0 high
	4K		167 00	2.960	2.235	0.725	72.5	0.1	72.4	97.6	247.7	Stream junction. "
	4L		140 00	2.845	2.345	0.500	50	0.1	49.9	98.1	248.2	Stream "
	4M		121 10	2.940	2.280	0.630	63	0.1	62.9	98.5	248.6	"
	4N		107 40	2.080	1.110	0.970	97	—	97.0	99.1	249.2	"
	4O		91 50	2.295	0.895	1.400	140	—	140.0	100.0	250.1	"
	ΔQ		223 03 20									

Fig. 45   Field book entries for a stadi-altimeter traverse.

The main traverse runs ΔP–Peg 1–Peg 2–Peg 3–ΔQ. Horizontal angles can be measured only on Face Left, because of the fitting of the additional eyepiece on the telescope. At ΔP the initial azimuth *PO* is known; this value is set on the circle, and ΔO is intersected using the *lower* plate controls so that the circle reading is unaltered; once the intersection has been made, the lower plate is left undisturbed and the *upper* plate controls are used to sight to points of detail and to Peg 1; the circle readings are therefore true bearings. The traverse leg from ΔP to Peg 1 is taped, with corrections for Abney slope readings. At Peg 1 the back bearing to ΔP is set on the circle by rotating the upper plate through 180° and clamping; intersection of ΔP is carried out by the lower plate controls.

That part of the traverse shown here illustrates detail mapped from Peg 3. For each point of detail the slant range is obtained, the slope correction is read directly from the graph, and the height scale is read to obtain the difference in height between Peg 3 and the detail observed. Plotting can be done by direct use of the circle readings as true bearings. Note that at Peg 3 the final circle reading is to Peg 4 to give the forward bearing over the next leg of the traverse. Before sighting to Peg 4 a check reading is usually taken on Peg 2 to ensure that the circle has not moved during the fixation of detail.

The distance Peg 3–Peg 4 could be read with the stadi-altimeter, but is preferably taped for greater accuracy of the main traverse line. The tacheometric readings from each peg are made with the main crosswire on a tape set on the staff at telescope height above the ground, the tape being set afresh with each move of the theodolite to a new station. If vegetation obscures the lower stadia line the horizontal crosswire is moved to 0.5 or 1.0 m above the tape and a note made in the Remarks column; the spot height entered will then need to be adjusted downward for plotting on the plane table or the drawn map.

At the final traverse station ΔQ the circle reading to ΔP, ΔR or some other known point should agree closely with the azimuth previously obtained from the control survey over the same line. Traverse bearings can be adjusted for misclosure as with a normal theodolite traverse if the misclosure is within acceptable limits. Details can be plotted directly from the circle readings as any adjustments to the bearings would be too small to affect protractor alignment.

graph. The eyepiece includes a prism which enables the observer to read the graph through a glass panel on the inner side of the box. The drum can be rotated by a milled disc at the observer's end of the box, this disc being graduated from 0 to 99 for reading against an index mark.

To adjust the stadi-altimeter, the telescope is set horizontal, i.e. the vertical circle micrometer reads zero with the sensitive bubble at centre. The altimeter eyepiece then shows a dot of light close to the vertical axis of the graph. The dot is brought on to the axis with an adjusting screw on the eyepiece bracket, and the drum is rotated so that the origin of the graph lies under the dot. The milled disc should read zero, and can be brought to this figure by rotating the disc while holding stationary the distant end of the drum.

Sights are taken to Scotch tape at the appropriate height on a levelling staff held at the point of detail required. As in the procedure described above, upper and lower stadia readings are booked (fig. 45) and checked for interval from the centre of the tape. The staff holder is then signalled to move to the next point. While he moves, the observer calculates the slant range and turns the drum until one-tenth of the slant range value, read from the black lines of the graph, lies under the light dot. The dot is then read against the red lines on the graph, the value obtained being the correction to bring the

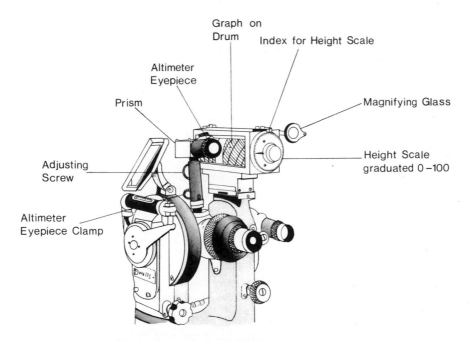

Fig. 46   The Ewing stadi-altimeter mounted on a theodolite. The optical eyepiece is fitted above the telescope. The graph can be seen in the window facing inwards towards this eyepiece. The height scale is visible on the end of the drum nearer to the observer: it is read against the small index mark attached at the top of the rectangular fitting. The metal pointer on the height scale disc should be ignored.

slant range to the true horizontal distance. The milled disc graduation opposite the index gives a reading of the difference in height between the
theodolite station and the staff station, to a maximum of 100 m difference.
The horizontal circle reading can be estimated to the nearest 5 minutes of arc
without moving the micrometer, if it is desired to book this. In the example
shown in fig. 44, the upper and lower stadia readings may have been 4.68 m
and 0.46 m; the slant range is 422 m; the drum is rotated to place the dot
at 42.2 on the black lines; the red lines read 11.3; the horizontal distance is
$422-11.3 = 410.5$ m to the nearest half metre. This instrument avoids
the use of tables and gives direct measurements. It can give direct readings
of spot heights if the milled disc is set not at zero but at the height of the
theodolite station, e.g. if the station is at an altitude of 242.2 m, the disc
is set at 42.2 with the telescope horizontal and the graph axis under the dot
of light. If the observation to the staff position gives a disc reading of 76.7,
the altitude of the staff base is either 276.7 or 176.7, and there should be
no doubt as to which is the correct value.

This instrument in effect gives a direct reading of the values otherwise
obtained from the stadia tables. There are, however, a large number of
other instruments which give more or less direct optical measurements of
distance, but most of these are specialist instruments. The non-specialist
will probably restrict himself to the use of the theodolite stadia lines in conjunction with a levelling staff, as described above, to the stadi-altimeter
attachment to the theodolite, or to the telescopic alidade, described on
p. 85, used in much the same way for stadia measurement.

Angular measurement with the sextant

The surveyor working on his own with no assistance, or with minimal
assistance, may prefer the use of a sextant to that of a theodolite. Its
advantages over the theodolite are

(a) weight: the theodolite and tripod may weigh over 10 lbs, compared
with 3 lbs for a sextant, a difference which is appreciable if one is
climbing steep slopes or if one has to carry equipment over a considerable distance.
(b) simplicity: the sextant has only one moving part and one control
knob.

The design, controls and theory of the sextant are given on pages 145–53.
The important points regarding its use are as follows:

1. vertical angles are not easy to measure if the sea horizon is not visible.
2. angles cannot be measured directly if larger than 125°, but can be
built up by combining angles each less than 125°.
3. horizontal angles need a tilt correction depending on the relative
elevation or depression of rays to the objects sighted.

4. a parallax correction must be made to horizontal angles if the R.O. used is nearer than 1 km.
5. angular measurements are usually taken to the nearest 12″ of arc.

Vertical angles

Usually the sea horizon will not be visible. It is possible to use an artificial horizon, but these are not always readily available. (Essentially an artificial horizon is a small dish containing a free mercury surface which will act as a horizontal reflecting mirror). The non-specialist will usually find it easier to measure vertical angles to the nearest 5′ with an Abney clinometer, or to use levelling for precise altitude work. If it is essential to take a vertical measurement with an accuracy greater than 5′ of arc, the sextant can be used as follows:

(a) Choose a clearly defined horizontal mark in line with the object to which a vertical angle is required, and at approximately the same altitude as the observer. Take a vertical reading to this mark with an Abney clinometer, measured carefully to the nearest 5′.

(b) Invert the clinometer and read again to the same mark. The angular measure should be the same, although read on the opposite side of the zero on the main scale. Repeat until one has five readings with the clinometer upright and five inverted, and take the mean value for the upright position, e.g.

Upright	E. 01° 10′	Inverted	01° 25′
	01° 15′		01° 20′
	01° 15′		01° 25′
	01° 10′		01° 20′
	01° 05′		01° 20′

The mean value accepted would be E. 01° 16′.5

(c) This is accepted as a reference level, and the sextant is then used to measure the vertical angle above this, to the nearest 0.2 minutes. The accuracy will not be to the nearest 12 seconds, of course, as the reference plane is defined only in terms of the accuracy obtained with the Abney clinometer, but it will be better than a single Abney reading. With a minimum of ten Abney readings as described one may assume that the reference plane is defined to an accuracy of at least one minute, and the final vertical angle (the sextant angle ± the Abney angle) will have this dependability.

To measure the vertical angle, the sextant is held upright. The reference mark is viewed directly through the telescope, and the movable arm is turned until the half mirror shows the image of the point whose altitude is required. The arm is then clamped and the image brought alongside the reference mark by turning the micrometer. The micrometer head is then read to the nearest 0.2 minutes.

Horizontal angles

The sextant is turned on its side to measure an angle which we can call
$A\hat{O}B$. Through the telescope A is viewed directly, and the movable arm is
turned until the half mirror shows the image of B. Final coincidence of A
and B is completed by rotation of the micrometer.

Measurement of the angle is thus very simple and very quick, particularly
as a modern sextant usually has a spring clamp on the movable arm which
holds it firmly to the scale unless it is squeezed, i.e. a firm grip on the arm
base releases the clamp and allows it to be swung rapidly to approximate
coincidence of A and B, the grip is then released and the micrometer screw
is used to complete the intersection.

The difficulty with measurement of horizontal angles lies in the elemination
of parallax on A and reduction of the measured angle to the true horizontal
angle. These two corrections are described on pages 149–50. Operational procedure is as follows:

Parallax   If A is less than 1 km from the observer, one should first sight to
A and obtain coincidence of direct and reflected images, reading the scale
to obtain the angle which is then to be applied, as a parallax correction, to
the observed angle after reduction to the horizontal.

Tilt correction   The angle measured is on an inclined plane. To reduce it
to the horizontal it is necessary to know the vertical angles to A and B.
Given these two elevations and the included angle it is possible to take
from table 7 the correction to be applied, interpolating as necessary. The
vertical elevations can be taken with Abney readings to the nearest 10′ (or
5′ if possible).

It must be appreciated that whereas a theodolite can be centred accurately
over a ground mark to within one or two millimetres, the sextant is held by
the observer standing over the mark. A displacement of 0.25 m is therefore
quite possible, and it is safer to assume a displacement of 0.5 m in assessing
accuracy of the survey. This means that the plottable error of the final map
must not be less than 0.5 m, which gives a scale of 1:2000 as the maximum.
It also means that observations must be to an accuracy such that any displacement resulting from approximation of angle must not be plottable
on the scale to be used.

A displacement of 0.5 m will result from an angular change as follows:

at range	1 km	from an angular change of	01′ 43″
	500 m		03′ 27″
	200 m		08′ 33″
	100 m		17′ 12″

Put in different form, an angular change of 1′ will result in displacements of

0.29 m	at	1 km
0.15 m		500 m
0.06 m		200 m
0.03 m		100 m

To be used in mapping on a scale of 1: 2000, therefore, horizontal angles measured with the sextant must be accurate to 01′ 43″ at a range of 1 km but can be less accurate at shorter ranges, as shown. Table 7 is therefore marked to show accuracy limits of 01′ 45″, 03′ 30″ and 08′ 30″ assuming that vertical angles have been measured with an Abney clinometer to the nearest 10′ of arc. The range of B will determine the accuracy limit required. If the horizontal and vertical angles involved take the correction outside the appropriate limit it is necessary to be more precise in observation and calculation. Greater accuracy in vertical angles can be obtained by using the sextant and a reference plane set out by multiple Abney readings, as stated above. Greater accuracy in calculation can be obtained by using the formula

$$\cos Y = \frac{\cos X - \sin b_A \sin b_B}{\cos b_A \cos b_B}$$

where Y is the horizontal angle required
X is the angle observed
b_A and b_B are the vertical angles of elevation to A and B, given as negative for angles of depression. The angles b_A and b_B are interchangeable in using the table.

The precise calculation may be done by using logarithmic tables or by using a calculator with keys for trigonometrical functions.

The most effective method of improving accuracy is to select as R.O. a clear, distant object at approximately the same altitude as the observer, and to measure the angle between this and A and also the angle between this and B. The two horizontal angles can be calculated with Table 7 and their difference used for the survey. An example indicates the merit of this procedure.

The survey requires the angle $A\hat{O}B$ to be measured, A and B being each 500 metres from the observer at O. The angle of elevation from O to A is +06° 00′ and from O to B is −05° 00′. The angle $A\hat{O}B$ measured with the sextant is 30° 00′ 00″. (Very simple figures have been used.) Table 7 gives no correction value corresponding to the three angles, indicating that the rate of change of the correction has become unacceptably large and the likely accuracy of the result correspondingly low. A calculation with the formula shows the correction to be −02° 02′ 43″, which would give a horizontal angle of 27° 57′ 17″. However, an error in measuring the elevation of A would be serious, e.g. if the true value were not +06° 00′

but were +06° 10′, the correction would be altered by 10′ 46″. This would cause a displacement by 1.6 m, which would be plottable on the scale of 1:2000. To improve the quality of the survey one has recourse to a point P, which has zero elevation from 0 and which gives values for the angles $P\hat{O}A$ and $P\hat{O}B$ of over 50°: the observed values are $P\hat{O}A = 60°\ 00′\ 00″$ and $P\hat{O}B = 87°\ 46′\ 00″$. Corrections can now be taken from the table, being $-00°\ 10′\ 57″$ for $P\hat{O}A$ and $-00°\ 00′\ 31″$ for $P\hat{O}B$. $A\hat{O}B = P\hat{O}B - P\hat{O}A = 87°\ 45′\ 29″ - 59°\ 49′\ 03″ = 27°\ 56′\ 26″$. This will be a more accurate result than the direct measurement. Furthermore, a 10′ error in measuring the elevation of B would now make a difference of only 38″, which would not give any plottable displacement.

The sextant therefore needs to be used intelligantly, but if handled correctly is very fast and has the great advantages of simplicity and of light weight. It could be used much more than it is for land surveys that do not need high precision. For offshore work it is of course the only practical instrument to use. Boat positions can be fixed by trigonometrical resection (see appendix I) with angles measured with the sextant.

Plane table traverses

Large scale

The plane table traverse is a technique admirable for many mapping projects, and too often ignored. The plane table sheet is first marked with the positions of the triangulation points already calculated. The plane table is then established on one of the fixed points e.g. P, and oriented on another by setting the alidade along the line between the two points on the map, and turning the table until the second point is intersected by the foresight of the alidade, when the table is clamped. A ray is then drawn towards a pole held on Peg 1. This ray, being part of the traverse line, may be drawn over its whole length from P to the estimated position of Peg 1. If the line is short on the map, as is often the case, a short extension mark may be made at the far end of the alidade, and marked as "P to Peg 1". Detail around P can be added by ray and distance methods, i.e. the assistant takes a pole and the zero end of the linen tape: the observer draws a ray to the pole and scales off the distance on the map, reading the tape at the station mark. A small number of rays may also be drawn to prominent features not yet on the map, e.g. a single tree on a hill-top, the junction of two paths or two streams, etc. None of these detail rays should be drawn all the way from the point P: to do so is to obscure the map unnecessarily. Figures 47 and 48 show correct and incorrect rays from the point P; the former are drawn only for short distances in the estimated positions of the detail, the latter obscure P, will have to be erased later, and over most of their length serve no useful purpose. The measured lengths will not conveniently exceed 60 m (i.e. two tape lengths) and slope corrections will

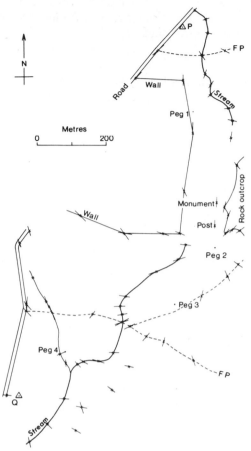

Fig. 47   The line of work from *P* to *Q* was plotted as a plane-table traverse: the station points could equally well have been plotted from theodolite traverse co-ordinates. From each traverse station rays have been drawn with the alidade to points of detail, but only a short length of ray in each case in the approximate location of the point. Distance (measured directly with a tape, or optically with stadia readings) is then set out to determine the precise location. By using only short lengths of ray the map is kept clear.

A similar map could be made from the stadi-altimeter traverse previously described.

then not be necessary unless the slope is greater than 03° on a scale of 1:1000 or 06° on a scale of 1:5000, as the correction is less than the plottable error, so that the measured lengths can be scaled directly on to the map. Before leaving *P* all detail within about 60 m should have been plotted for position. Height information should also be added using the Indian clinometer. A piece of Scotch tape can be attached to the assistant's pole at a height above ground equal to that of the backsight of the clinometer on the table. The observer sights to the pole at each point of detail, brings to centre the clinometer bubble, brings the foresight in line with the tape, and reads the Natural Tangent scale. The difference in altitude between the plane table site and the pole position is Δh, and

Δh = natural tangent reading x horizontal distance.

This can be calculated immediately, and a spot height written on the map (the altitude of point P being known from the triangulation). If formlines

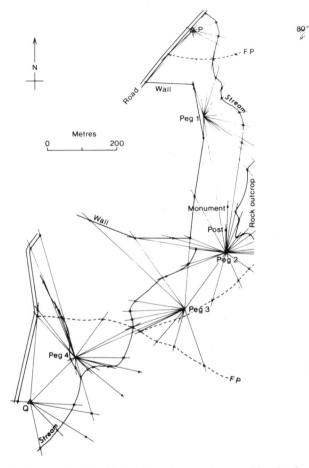

Fig. 48   As for fig. 47, but here the map is marred by the drawing of rays radiating from the observation points. These rays obscure the map and are unnecessary: all that is required is a short length of each in the vicinity of the detail to be plotted.

are sufficient for the purpose of the survey, formlines around P can be interpolated between spot heights, before the table is moved. A tangent reading to Peg 1 is also noted.

The line from P to Peg 1 is chained in the normal manner. It is desirable, as a precaution against error, to write down in a field book each chain length and slope reading as measured with the Abney clinometer. Corrections for slope are entered at once from table 3. Offsets and points of on-line detail need not be entered in the field book. Instead, such detail is drawn directly on the map, scaled from P towards Peg 1 and offset at right angles as necessary. When the chaining to Peg 1 is completed and the correct distance calculated, the position of Peg 1 can be scaled from P and plotted on the ray already drawn to it.

The plane table is now established over Peg 1, levelled, and the alidade is set along the line P to Peg 1 previously drawn, but with the foresight towards P. The table is then rotated until the foresight intersects the signal at P, when the table is clamped. As with all plane table sights, it is important after clamping to check that the foresight is still on the target: this should be done without touching any of the equipment, so that there is no possibility of accidental sideways pressure, from the observer's hand, which when released might allow slight movement of the table.

The table is now oriented, and a forward ray is drawn to the pole at Peg 2. As before, this may be drawn over its whole length, being part of the traverse line. Tangent readings are taken with the clinometer to both P and Peg 2. The altitude of Peg 1 can now be calculated, by multiplying the distance (P to Peg 1) by the mean value of the two tangents observed over the line: the product is the difference in altitude, and this is applied to the altitude of P, with correct sign, to obtain the altitude of Peg 1.

Detail around Peg 1 is then mapped as before, by ray and taped distance, and height information again drawn in. As before, rays to a few prominent marks may be drawn, provided that in each case the angle of intersection with the ray from P is not less than $30°$—if the angle is more acute, there is no value in drawing the ray from Peg 1. When detail mapping is complete, the line Peg 1 to Peg 2 is chained and plotted, and the table is positioned and oriented at Peg 2. The traverse proceeds in this manner until a final ray is drawn to the fixed point on which the traverse is to close: this ray should pass through the point marked on the map, and the measured distance should exactly match the marked point. If the work has been done with maximum care, there should be no misclosure. If a misclosure is found, there is no means of rectifying it, and the traverse and its attendant detail must be deleted back to the last point which can be guaranteed as correct. For this reason it is highly desirable that at any traverse point from which triangulation stations can be seen, back rays from these stations should be drawn after orientation of the table on the previous traverse point, as a check on the accuracy of the work thus far. In other words, a form of resection check (see p. 90) is put on the traverse whenever possible, and this is particularly necessary if the traverse includes short legs over which the possibility of slight angular error in orientation is most likely.

It must be remembered that on a scale of 1:1000 the plottable error is 0.250 m. The table should therefore be set up over the station mark in such a position that it is the traverse point as drawn, and not the centre of the tripod, which is immediately above the mark. In this way it is possible to avoid small errors of orientation which result from mis-alignment. The forward ray was drawn to the station mark, and reverse alignment must be from the same point and not from one a little to one side. This careful centring over the station mark of the plotted position of the actual working point is less important as the scale decreases; it increases in importance when either of the traverse legs from the station is short.

This form of traverse achieves a great deal very quickly, if properly executed. The traverse pegs are fixed points which can be used for subsequent work, and an irregular belt of detail accompanies the traverse line as the latter is extended. The rays to prominent features fix these (provided that three rays intersect at a point at angles between 30° and 150°) thus providing additional points for later use in the survey. For close detail this is the best survey technique available, e.g. the mapping of an unplanned African village with a scattered and irregular collection of round huts, grain stores, compound walls, etc., is probably practicable only with this technique (excluding the use of aerial photographs and photogrammetric plotters); similarly the mapping of bed deposits in an incised river, with banks from 5 to 10 m in height. The traverse should never be left "open", in the sense that its final peg positions are unchecked for location, as clearly any error in the alignment of a forward ray will swing out of their true positions and bearings all the subsequent sections of the traverse. It is, however, permissible to terminate the traverse at a point checked by resection from previously surveyed stations.

All plane table drawing should be done with a hard pencil, such as 4H quality, which will give a clear, firm line of precise location if the point is kept sharpened. A soft pencil not only gives a blurred line of indistinct location, but rapidly causes the whole map to deteriorate, as movement of the alidade smears the pencil lines.

Small scale

The basic principles remain unchanged, but the difference in plottable error introduces much greater freedom of movement. For example, mapping on a scale of 1:10000 gives a plottable error of 2.5 m. Slope corrections, except in very steep terrain, can be ignored, and, indeed, measurements to points of detail can be made by pacing within the limits of accuracy prescribed. On this scale it is possible for the observer to work alone if it is absolutely essential, pacing even the traverse legs, provided that he knows from past experience that he can maintain a regular length of pace and provided that every second or third peg position can be checked by resection rays. At this scale the plane table traverse becomes a half-way stage towards survey by successive resections, but is preferable to the latter in country partially obscured by vegetation.

Plane table traverses with tacheometric measurement

It is possible to perform a plane table traverse with length measurement or detail fixation achieved optically in much the same way as with a theodolite traverse. The essential item of equipment is a telescopic alidade with stadia lines. This is a relatively heavy accessory to the plane table, and should be employed only when the table tripod is substantial and provides a firm foundation for work. It is not really suitable on a light reconnaissance tripod with telescopic legs.

The telescopic alidade in its basic form consists of a telescope, mounted on an alidade base, with a vertical arc allowing the reading of angles of elevation and depression. The telescope has clamp and tangent screws, focussing screw, etc., and the method of reading the vertical arc may be by a vernier or by a micrometer. As with the theodolite, observation to a graduated staff allows determination of horizontal distance and of difference in altitude. The recommended length of Scotch tape is again attached to the graduated staff before observations start, but this time at a lower level equal to the height above ground of the alidade telescope. As bushes, bracken, etc., may obscure the line of sight, a second piece of tape should also be attached, preferably 2.000 m above the first piece. Distance to the staff is then found as described on p. 68, i.e. the sensitive bubble on the alidade is brought to centre, the horizontal crosswire is made to intersect the Scotch tape, the upper and lower stadia readings are recorded, the intervals of these from the crosswire are checked for rough equality, and the slant range is calculated. The angle of elevation or depression is read from the vertical arc, and this angle (which need be read or estimated only to the nearest 10 minutes) is used in conjunction with the slant range to extract from tables 4A and 4B the true horizontal distance to the staff and the difference in altitude between the plane table position and the staff position. Spot heights are therefore put on the map as before, and formlines can be interpolated in the field while looking at the terrain. If observation has to be made to the upper length of Scotch tape, 2.000 m will be subtracted from the calculated height of the staff base. The traverse legs can be similarly measured if scale permits. At points from which triangulation stations are visible, a check can be made not only on the location of the traverse point but also on the height of the point. The latter is calculated either from readings to the staff held at the triangulation station, if this is sufficiently close, or by observing the vertical angle to approximately the correct height above ground on the triangulation target and then multiplying the horizontal distance as measured from the map by the tangent of the vertical angle, as given in table 5; either method gives the difference in altitude between the triangulation station and the traverse point. If several triangulation stations are visible, it is desirable to observe to all and to calculate the traverse point altitude from each. It should be possible to obtain three altitude values agreeing within a metre.

Direct-reading tacheometric instruments

The Beaman arc alidade produced by Hilger & Watts (Rank) has an additional eyepiece, mounted on the telescope, through which three scales can be viewed (fig. 49). The recommended procedure with the instrument is as follows:

(1) Point the alidade at the staff held at the detail to be mapped, and bring the bubble of the main spirit level to centre with the levelling screw.
(2) Sight to the staff at about the 2 m mark. Then look in the small eyepiece and raise or lower the telescope by means of the vertical tangent screw until a graduation on the *V* scale coincides with the index line.
(3) Record the *V* reading (whole number) and the *H* reading (to one decimal place).
(4) Read and record the staff graduations under the central crosswire, the upper stadia and the lower stadia.

The horizontal distance and the altitude difference are then calculated as shown on the form in fig. 50. Essentially the instrument gives a direct reading of $100 . \sin \theta . \cos \theta$ (the *V* number) and of $100 . \sin^2 \theta$ (the *H* number) and the basic calculation is as before.

(a)

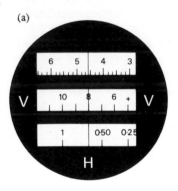

Fig. 49   The eyepiece of the Beaman arc shows parts of three scales, as shown in (a). (b) illustrates sections of the three scales over a greater range, with factors *V* and *H* set against the corresponding vertical angle.
    If desired, the instrument can be used without reference to the *H* and *V* factors, reading only the vertical angle to the nearest 10 minutes and employing the tacheometric tables and stadia readings.

(b)

Vertical Angle

| 0° | 1° | 2° | 3° | 4° | 5° | 6° | 7° | 8° | 9° | 10° | 11° | 12° | 13° | 14° | 15° |

Horizontal Factor = $100 \sin^2 \theta$

0   0·1   0·2 0·3 0·4 0·5   1·0   1·5   2·0   3·0   4·0   5·0   6·0

Vertical Factor = $100 \sin \theta \cos \theta$

0 1 2 3 4 5 6 7 8 9 10 11 12 13 14 15 16 17 18 19 20 21 22 23 24 25 26

Survey: *Plane Table Traverse P–Q* Observer: *C.D. Jones.*

Station: *Pt 2* Date: *18.8.73*

Altitude of Station: *268.0 metres* Height of Instrument (i): *1.00 metres*

BEAMAN ARC

TO	V (insert sign)	H	STAFF READING a	UPPER STADIA b	LOWER STADIA c	STADIA INTERCEPT (b-c)	SLANT RANGE 100(b-c) =d	CORRECTION H(b-c)=e	HORIZONTAL DISTANCE d-e	VERTICAL COMPONENT f=V(b-c) (insert sign)	CORRECTION IN HEIGHT +i-a (insert sign)	DIFFERENCE IN ALTITUDE f+i-a (insert sign)	ALTITUDE
2A	-9.0	0.8	1.030	1.630	0.430	1.200	120.0	0.96	119.0	-10.80	+0.07	-10.73	257.27
2B	-7.0	0.5	1.310	1.815	0.805	1.010	101.0	0.51	100.5	-7.07	-0.21	-7.28	260.72
2C	-5.0	0.3	0.780	1.020	0.565	0.475	47.5	0.1425	47.4	-2.38	+0.32	-2.06	265.94
2D	-2.0	0.1	0.980	1.345	0.620	0.725	72.5	0.0725	72.4	-1.45	+0.12	-1.33	266.67
2E	Level	0	1.100	1.540	0.660	0.880	88.0	0.0	88.0	0.0	-0.00	0.00	268.00
2F	+2.0	0.1	1.900	2.375	1.420	0.955	95.5	0.0955	95.4	+1.91	-0.80	+1.11	269.11
2G	+2.0	0.1	1.990	2.795	1.190	1.605	160.5	0.1605	160.3	+3.21	-0.89	+2.32	270.32
2H	+2.6	0.1	1.100	1.890	0.295	1.395	159.5	0.1595	159.3	+4.15	nil	+4.15	272.15
2J	+2.6	0.1	"	1.650	0.550	1.100	110.0	0.110	109.9	+2.86	"	+2.86	270.86
2K	+3.5	0.1	"	1.555	0.650	0.905	90.5	0.0905	90.4	+3.17	"	+3.17	271.17
2L	+0.9	0.0	"	1.540	0.655	0.885	88.5	0.000	88.5	+0.80	"	+0.80	268.80
2M	-1.4	0.0	"	1.800	0.400	1.400	140.0	0.000	140.0	-1.96	"	-1.96	266.04
2N	-7.2	0.6	"	2.050	0.150	1.900	190.0	1.140	188.9	-13.68	"	-13.68	254.32

Fig. 50 The calculation of observations with the Beaman Arc entails considerable arithmetic. The form illustrated is largely self-explanatory, and is intended to establish a routine for the calculation. The data for the observations to points 2A–2G were made with the V scale on a whole number. This simplifies one multiplication but introduces the necessity of a correction for staff reading and for instrument height, with the possible introduction of errors of sign. The observations to 2H–2N were made with a tape on the staff at a height above ground equal to that of the instrument telescope, when V and H are both estimated to one place of decimals. This slightly complicates the calculation of f but removes the correction.

The simplest method is to avoid all use of the Beaman Arc and to employ the instrument as a simple alidade, reading the vertical angle and the stadia values in conjunction with the tacheometric tables.

It is, however, exceedingly tedious to use, and opportunities abound for arithmetical error. The form shown is intended to simplify the calculation, but even so cannot avoid arithmetic such as for point 2C:

$(1.020 - 0.545) \cdot 0.3 = 0.1425$

$(1.020 - 0.545) \cdot -5.0 + 1.100 - 0.780 = -2.06$

which most field workers will wish to do with pencil and paper or with a slide rule rather than relying on mental calculation.

It is easier to observe to a piece of tape on the staff at a height above ground equal to that of the alidade telescope axis. This avoids the correction $(+ i - a)$ where i is the telescope height above ground and a is the staff reading from the horizontal crosswire, but it introduces into V a figure in the first place of decimals to give, for example in the case of $2H$:

$+2.6 \cdot 1.595 = 4.15$

again needing a paper calculation. On the whole, this method with the tape slightly reduces chances of error, but it is far easier to use the Upper and Lower Stadia readings with the tacheometric tables, ignoring V and H readings and instead taking the vertical angle, to the nearest 10 minutes, from the eyepiece scale, with the crosswire on the tape in a manner similar to that of theodolite stadia readings.

It is possible to obtain telescopic alidades with special features to give direct readings for horizontal distance and altitude difference. The design varies very widely. Some use a horizontal bar instead of a vertical staff. Many European manufacturers of tacheometers also produce alidades fitted with optical systems of comparable design.

All telescopic alidades can be used in much the same way as the tacheometric theodolite, except that the bearings are not recorded, the rays being drawn with the alidade and the points plotted to scale as they are observed. As a result there is a saving of time, and contours can be drawn in the field instead of in the laboratory. There is the usual advantage of the plane table that the map can be inspected for omissions at each point in the field, before a move is made to the next station. The alidade cannot be used in rain without an assistant holding an umbrella over the plane table, an arrangement which is rarely satisfactory, and the map cannot be enlarged without loss of accuracy. The theodolite map can, in contrast, be replotted at any time and on any scale up to 1:1000 without loss of accuracy.

3 Infilling of further detail

Plane table resections

The previous chapter included an account of the use of the plane table for traverse purposes, when it is established at a point already fixed, is oriented on a second point already plotted, and is then employed in the fixation of a line to a new traverse point to which it will subsequently be transferred for back orientation using the same line.

In resection techniques, the plane table is taken to a point whose position is as yet unknown, and the first operation is to determine the position and orientation, after which surrounding detail can be mapped by ray and distance methods. There are three methods of full resection and one of partial resection.

Partial resection

This is the simplest method. In the course of the survey it may be realised that detail needs to be mapped in an area not yet visited and in which there are no trigonometrical stations and no traverse points. It would be possible to run a plane table traverse into the area from a fixed point, but a semi-resection is far quicker. If the table is already at point P, where it has been oriented and where detail has been surveyed, the assistant can set up a pole at the new point at which it is desired to work. A ray is drawn from P to the pole, and the plane table can then be transferred to the new point, centred and levelled. The alidade is then laid along the line just drawn, but facing in the opposite direction, and the table is rotated until the foresight intersects the signal re-established at P. The table is then clamped, and it is obviously correctly oriented. The position of the new point is not yet known, but can be readily obtained if two other fixed points Q and R are visible. The alidade is pivoted around the marked position q on the table, the Q signal is intersected with the foresight, and a ray is drawn back from the mark q to intersect the orientation ray from P. This is the observer's position. A similar sighting is made on R, and a ray drawn back from the plotted position r, and this should confirm the first intersection by passing through the same point. Mapping of detail around the new point can then proceed.

Full resection. Method I (Bessel's)

Of the full resection methods, Bessel's is the quickest and easiest. In fig. 51 let the points A, B and C be three of the triangulation stations which have been plotted on the plane table as a, b and c. The observer establishes the plane table at a point P which he wishes to fix preparatory to local mapping of detail. Having levelled the table, he sets the alidade along the line ab and turns the table until he intersects B with the foresight. He clamps the table, pivots the alidade round point a, and having sighted to C draws a ray ac. The table is then unclamped, the alidade is reversed so that it lies along the line ba, and the table is turned until A is intersected, when the table is again clamped. The observer pivots the alidade around b, sights to C and draws a

a) Location of points

$B_\triangle$ $_\triangle C$

$A_\triangle$ $_\times P$

b) Enlargement of plane table plot

Fig. 51   Bessel's method of plane table resection. See text for the successive steps. The two arrows in (b) represent the rays ac and bc, drawn towards C when the board was turned into its first two positions. The ray bc has been extended backwards to meet ac at d.

   The broken lines are the rays drawn back from b and c after orientation of the board by alignment of dc with C. The point p is the mapped location of the observer's position P.

ray *bc*. The rays *ac* and *bc* are extended if necessary and their point of intersection can be called *d*. The alidade is then laid along the line *dc*, the table is unclamped and is turned until the alidade foresight intersects *C*, and the table is then re-clamped. It is now correctly oriented. As with the semi-resection above, the position of the observer is now found by pivoting the alidade round *b*, sighting *B* and drawing a ray back from *b* to cut the line *cd* at *p*. A similar back ray from *A* should also pass through *p*, thus confirming the position. The proof of this method is given on page 189.

Full resection. Method II (tracing paper)

A sheet of tracing paper is laid on the table at *P*, no attempt being made to orient the table, which is clamped once it has been levelled. From any arbitrary point on the tracing paper a ray is drawn towards *A*, which is sighted with the alidade. A second ray is drawn towards *B* and a third towards *C*, all three rays radiating from the same point on the tracing paper. This paper is now moved over the plane table until the ray to *A* passes through the plotted position *a*, the ray to *B* passes through *b* and the ray to *C* passes through *c*. The common point from which the three rays radiate is now at the observer's position, and may be marked on the plane table map. This transference of the point should never be done by "pricking through" with a pin, which mutilates both the map and the table: a soft pencil should be used to shade the underside of the tracing paper at the radial point, and when the sheet is in position a small cross stroke at the point will transfer the position to the map. The alidade is then laid from this position to that of the most distant of the visible triangulation points, and the table is turned until the foresight intersects the station target, when the table is clamped. The orientation is checked by laying the alidade from the newly marked point to another fixed point already plotted, and sighting through the alidade to confirm that it is also on line with this check station.

The disadvantage of this method is the need to transport a large sheet of tracing paper. A cover sheet of tracing paper can be useful protection against scattered raindrops and can be pinned on the underside of the table, being folded across the map when the table is being carried, but folded out of the way when a station is being worked. Such a sheet can be used for resection purposes provided that it is cleaned on both sides after each resection.

Full resection. Method III (triangle of error)

The plane table is established and levelled at the unknown point *P*. At a previous position, when correctly oriented, a line should have been drawn along the side of a trough compass laid on the table with the needle at the centre graduation. At *P* the compass is again laid alongside this line, and the table is turned until the needle is once more at the centre graduation. The table is then clamped, although orientation is not yet complete. Local attraction may deflect the compass needle by several degrees, so further orientation procedure is necessary. The compass is removed, and the alidade is pivoted around the point *a* and is sighted on *A*: a back ray is drawn from *a* towards the observer (this need be drawn only in the area of the map in which the point *p* can be expected to lie). Similar rays are drawn back from

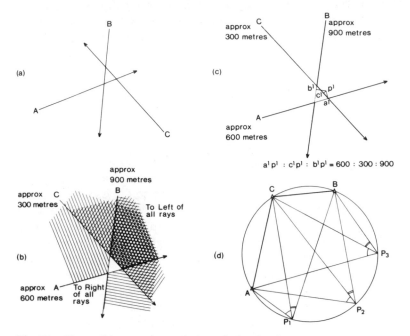

Fig. 52   Plane table resection: solution of triangle of error.
(a) Rays forming a triangle within which lies the desired point.
(b) Rays forming a triangle outside which lies the desired point, which will be either to the left or to the right of all rays.
(c) Location of trial point p^1 proportionately distant from all rays.
(d) The "Danger Circle" situation. The locating angles will be identical at all points on the circumference of the circle.

b and from *c*. It is unlikely that the three rays will intersect at a point: usually they form a small triangle known as the "triangle of error". This arises because the table is not yet correctly oriented, but requires to be rotated by a small angle to complete correct orientation. All rays drawn with the incorrect orientation will therefore be displaced laterally by the same small angle; the three rays drawn are thus displaced either all to the left or all to the right depending on the angular error of the table. Clearly, also, the amount of displacement of the incorrect line from its true location will increase with distance from the fixed point through which both pass. To solve the triangle of error to find the correct position of *p*, it is necessary to satisfy two conditions:

(i) all rays will be deflected either to the left or to the right, viewed from the respective fixed points;

(ii) the distance of the correct position p from each ray will be proportional to the length of the ray.

If the point P is inside the triangle formed by the three fixed points A, B and C, the required position on the plane table will be inside the triangle of error and can be found by applying condition (ii) (see fig. 52a). If the point P is outside the triangle ABC, the required position p is outside the triangle of error. Condition (i) is applied first, and this will identify two sectors in which the point may lie, one being to the left of all three rays and the other to the right (fig. 52b). Condition (ii) can be applied to only one of these sectors (fig. 52c) and an estimated true position is drawn at the appropriate point. All trial work is then erased, and a second attempt made at orientation. The alidade is laid along the line from the new estimated position to the fixed point which is most distant, and the table in unclamped and turned slightly until the alidade foresight intersects the station target, when the table is re-clamped. As before, rays are now drawn back from a, b and c. Probably a triangle of error is still present, but smaller (if it is larger, but of similar shape, the table was rotated in the wrong direction, probably as a result of the new estimated position being placed in the wrong sector; if smaller, and of similar shape, the table needs a little more rotation in the same direction as before; if the triangle has "turned over" to give a mirrored shape compared with the original, there has been over-correction and the table needs to be rotated in the opposite sense, by a large or small amount depending on the size of the triangle). The procedure is repeated until the three rays pass through one point, when the table is correctly oriented, its position is known, and local mapping of detail can be started.

As described here, the resection started with approximate alignment by the trough compass. This saves time, but can be dispensed with. If no compass is available, the observer starts by marking his estimated position, and the resection then continues as though this point had been found from the first triangle of error.

General comments

If the observer is on, or is nearly on, the circumference of the circle passing through A, B and C, no resection is possible. The "danger circle", as it is called, gives an infinite number of possible solutions to Method II and Method III. For this reason the observer should always be suspicious if Method III gives a near-perfect intersection of rays at the first attempt, and he should always test Method II for other possible locations. Method I becomes impossible under "danger circle" conditions, which is an additional point in its favour. If "danger circle" conditions appear to be at all possible with the three fixed points first selected, a new selection should be made with which point P cannot be concyclic.

Once the location of P has been determined, the altitude can also be found. The Indian clinometer is pointed at a station of known altitude, is levelled, and then has its foresight brought to rest at a point on the target pole which is the same height above the ground as is the backsight of the clinometer. The natural tangent is then read and is multiplied by the distance (measured from the map) to find the difference in altitude. This difference is applied to the known altitude of the target station, using the correct sign, plus if P is above the other station, minus if P is lower than the other station, to determine the altitude of P. Subsequently detail is added around the observer by ray and distance methods: distance can be taped or measured optically, and spot heights can be fixed for the insertion of formlines or contours. If the observer is using a telescopic alidade, he can obtain his altitude by a sighting to a station of known altitude as described on p. 85.

Contouring

The method of using the Indian clinometer to mark spot heights has already been described (p. 81), and in the course of detail survey with the plane table a sufficient number of spot heights can be inserted for good formlines to be interpolated with a high degree of accuracy. To be called a contour, such a line should be delimited by the fixation of a number of points at the contour altitude, the marked positions of these points then being joined together while the observer is studying the topography in the field.

The Indian clinometer can be used for contour surveys, in conjunction with a levelling staff. If the altitude of the plane table station is known (either because it is a traverse point or because its position and altitude have been fixed by resection) contours may be marked out around it. The levelling staff is stood alongside the plane table and the height of the clinometer backsight is noted. For example, let it be measured as 1.230 m. The altitude of the ground at the plane table site is known to be 186.420 m above sea-level. It is required to mark on the ground the 185-metre contour. A horizontal line of sight from the clinometer will be at 186.420 + 1.230 = 187.650 m. It is desired to find points at which the staff base will be at the desired altitude of 185.000 m, i.e. 2.650 m below the line of sight. A piece of Scotch tape is therefore attached to the staff across the 2.650 graduation. The assistant then stands approximately on the 185-metre contour, while the observer sights the clinometer at him. With the clinometer bubble brought to centre by the levelling screw, the observer sets the foresight at zero and studies the piece of tape in relation to the foresight. He directs the assistant upslope or downslope until the crosswire lies on the tape, when the assistant inserts a contour peg and moves to a new position at which the procedure is repeated. In due course the 185-metre contour will be marked on the ground by a line of pegs, and these can then be surveyed by ray and measured distance in the ordinary way.

If a telescopic alidade is available, this can be used in place of the Indian clinometer, with the sensitive bubble kept central in the spirit level and the telescope maintained in the horizontal position giving zero on the vertical circle. In this case the delimitation of the contour and its plotting on the map can be combined, and there is no need to set out pegs. Once the staff has been sited on the contour, a ray is drawn to it with the alidade, and the horizontal distance is measured tacheometrically by reading the upper and lower stadia. As the line of sight is horizontal, the distance equals 100 . (upper stadia − lower stadia).

The most accurate technique of contouring is, of course, with the use of a level. If two assistants are available, the combination of a quickset level and a plane table is rapid, accurate and effective. The quickset level is used to bring heights from the nearest accurately-determined station altitude, by the method described on p. 127, to the area in which detail is to be surveyed. The plane table is set up about 5 m from the level, the table position is resected, and the table is oriented. The position of the level is marked temporarily on the map, being fixed by ray and distance. If the telescope altitude of the quickset level is, for example, 137.110 m above sea-level, a reading of 2.110 m on a levelling staff would mean that the base of the staff was on the 135-metre contour. A piece of Scotch tape is therefore attached to the staff face across the 2.110 graduation, and the staff tested in different locations until the main crosswire of the level telescope lies on the tape, with the sensitive bubble in centre. The staff position is then plotted: a ray is drawn on the map from the plane table position to the staff, and the distance (upper stadia reading minus lower stadia reading, multiplied by 100, as the line of sight is horizontal) is scaled from the plotted position of the level to a point on the ray. In this way the contour can be inserted very rapidly. Note that it is not necessary to have the telescope crosswire exactly on the 2.110, graduation—anywhere on the tape will suffice. This is because the very minor lateral displacement required to locate the staff in the exact position will be indistinguishable from the approximate position on the scale used, and a grass tussock or a small stone can give vertical changes which have no locational significance. If contours are being inserted at a Vertical Interval of 1 m, then the 136-metre contour can be plotted by repeating the operation with the tape at 1.110 m, and the 137-metre contour with the tape at 0.110 m, although the last-named would clearly be unworkable in practice as the lower stadia would not fall on the staff. When all the required contour information has been plotted, other detail may be added, using the level as a tacheometer. Vegetation boundaries, for example, can be plotted by rays from the plane table mark and distances scaled from the mark of the level position. For this detail mapping no Scotch tape is required on the staff; the upper and lower stadia readings can be at any height on the staff provided that the sensitive spirit level shows that the telescope is horizontal. As a check on the distances, the usual precaution should be taken of checking the two intervals from the crosswire reading to the two stadia readings. For short lines, of course, it may be easier to make a direct measurement of distance from the table by means of a linen tape. In choosing contour positions the assistant should if possible attempt to locate contours where they cross paths or vegetation boundaries which are also required on the map.

Because the contour positions, although accurate in location, may vary fractionally in altitude because of the use of the tape instead of the precise reading on the staff, the levelling should never be carried forward from a contour point. When the survey has to be moved to a new position, the staff is located at any convenient point, the graduations are read as usual to .005 m, and the line of levels proceeds in orthodox fashion.

Theodolite in conjunction with the plane table

The theodolite can also be used as an accessory to the plane table. The latter is resected as usual, its position is marked on the map, and the table is oriented. The theodolite is set on its tripod about 5 metres distant, and its position is plotted on the map. If the telescope is set horizontally, with the vertical circle reading zero and the sensitive bubble central, the theodolite can be used for contouring in the same way as the quickset level described above. Because the theodolite does not have to be centred over a mark, it can be set up and levelled with ease. The initial altitude in this case is obtained by a vertical angle to a known station, and calculation of the altitude difference by multiplication of the distance, scaled from the map, by the natural tangent of the angle of elevation or depression (see table 5).

The theodolite is more versatile than the level by virtue of its vertical circle, and the levelling staff can be used as described on page 68 for fixing points of detail: rays are drawn from the plane table position and distances are scaled from the theodolite position, the theodolite being used essentially as a range finder, but with the added quality of also producing spot heights. If the theodolite is fitted with a stadi-altimeter or other direct-reading attachment, the mapping is correspondingly accelerated. It is not essential to have two assistants for this type of work, since the observer can if necessary operate both the plane table and the theodolite. Although this is inevitably slower than if two people are working the instruments simultaneously, it is still much faster than if all distances have to be taped. In addition, the range of mapping is extended to about 200 m.

This symbiosis of theodolite and plane table can be exercised both ways. As described here, the theodolite is an accessory to the table, but the table can equally well be used as an accessory to a theodolite traverse in which detail is being added tacheometrically. In that case the plane table is set up within 5 m of the theodolite station, and detail is plotted directly as the observer enters data in the field book. For the move from station to station, the plane table is set up briefly over the new station traverse peg, is oriented on the back station and a ray is drawn to the forward station, which is plotted when the distance has been chained. The site of the plane table, within 5 m, is also plotted before the table is moved there; orientation of the table at this site is achieved by sighting on to the previous traverse peg, as the peg now occupied by the theodolite is too close for reliable orientation.

Prismatic compass traverse

Traverses can be run with a prismatic compass. The most common use for a compass traverse is the plotting of a path, a stream, etc., in which the emphasis is on a long, narrow feature which needs to be added to the map. Although described here, it should be stressed that such traverses have a very low standard of accuracy, usually with misclosures of the order of 1:100 to 1:50, and other surveying techniques should be preferred if possible. It is possible to complete a compass traverse of 30 km in a day. Compass traverses are excellent for reconnaissance sketch maps where high accuracy is not desired, but their limitations must be appreciated.

The prismatic compass allows reading of magnetic bearings, measured clockwise from the magnetic meridian. Surveys based on magnetic bearings are best plotted in relation to magnetic north, but with true north added to the map, except where an addition is being made to a map which otherwise is entirely on true bearings or co-ordinates. In this case the magnetic bearings should be corrected to true bearings before plotting. The difference between Magnetic North and True North is the Magnetic Declination, expressed in degrees East or West of the true meridian. The Magnetic Declination varies from place to place with time. The annual change at a place is known as the Annual Magnetic Variation, and in making conversions from magnetic to true bearings it is important that the current Magnetic Declination should be used. For example, if the local Magnetic Declination is not known, but a 1961 map shows it as about $9\frac{1}{2}°$W., with annual variation about $7\frac{1}{2}'$E. per annum, a compass survey run in 1973 would have bearings corrected to true as follows:

1961 Magnetic Declination	9° 30′ W.
12-year interval = 12 × $7\frac{1}{2}'$	1° 30′ E.
1973 Magnetic Declination	8° 00′ W.

With a compass traverse, the bearing of each leg is measured with the prismatic compass, and the distance is chained (or paced, if the scale is small). The lines are normally chained in the usual way, with distances and offset data entered in the field book. Also entered are the forward and backward bearings of each leg, and bearings to any other points. As with true bearings, forward and backward magnetic bearings over a line should *differ* by 180°.

A forward bearing is taken from a known point such as P to the first traverse point Peg 1. The bearing should be read to the nearest $\frac{1}{2}°$, and if possible to the nearest $\frac{1}{4}°$. At Peg 1 a back bearing is taken to P. The back bearing should equal the forward bearing ±180°, but will quite possibly differ by some other figure, due to local attraction of the compass needle at one or both stations. As long as the difference lies between 178° and 182° the bearings can be accepted, the agreed forward bearing being taken as

$$\frac{\text{Forward Bearing} + (\text{Back Bearing} \pm 180°)}{2}$$

If the forward bearing is 327° and the back bearing is 145°, the accepted forward bearing used for plotting the traverse will be 326°; if the forward bearing is 27° and the back bearing 208°, the accepted forward bearing will be 27½°. If the two bearings do not agree within the 2° limit allowed, they must be repeated. If they still disagree, it is probable that one station suffers from marked local attraction, and the site may have to be altered slightly. If it appears impossible to eradicate the error, it may be necessary to calculate the location and extent of the attraction, and to compensate it. For example, assume the following readings are taken on a traverse:

At Station A	Forward Bearing AB = 67½°	
At Station B	Forward Bearing BC = 136°	Back Bearing BA = 247°
At Station C	Forward Bearing CD = 54°	Back Bearing CB = 311½°
At Station D	Forward Bearing DE = 39½°	Back Bearing DC = 239°
At Station E		Back Bearing ED = 219°

Over the line AB, the two bearings agree within ½°, and the forward bearing can be accepted as 67¼°. It appears that there is no unusual local attraction at either point. The forward bearing of 136° at B is therefore likely to be correct, but it differs widely from the back bearing at C, which is 311½° instead of the expected 316°. The forward bearing from C to D also differs from the back bearing from D beyond the tolerable limit. Over the line DE the forward bearing is 39½°, and the back bearing 219°, a difference of only ½° from expectation, so again it may be assumed that readings at both D and E are reliable; the forward bearing DE will be accepted as 39¼°. All the discrepancy therefore appears to be located at C. The back bearing to B is 4½° less than expected; the forward bearing to D is 5° less than might be expected. It may therefore be assumed that there is a deviation at C of 4½°–5°, due to local attraction. Both the bearings measured at C may therefore be increased by 5° to new values of 316½° and 59°, which can then be considered respectively with the reverse bearings of 136° and 239°. The accepted forward bearings are then 136¼° for BC and 59° for CD.

Methods of booking data vary. The chainage line measurements are entered as usual in the centre column, with offset data to left and right. Bearings can be entered as shown in fig. 53, the forward bearing being immediately above the traverse station name, and the back bearing immediately below the destination name. Each traverse leg should be started on a fresh page of the field book.

Traverse adjustment

Each leg of the traverse is plotted by setting out the bearing with a protractor from the north line. (The plotting of the traverse line is best done on squared paper and transferred to the map in its adjusted form before the addition of detail.) Any error in the bearing of one leg is therefore not cumulative, in that it does not "swing" all subsequent sections of the traverse in the way that an angular error does in a standard theodolite

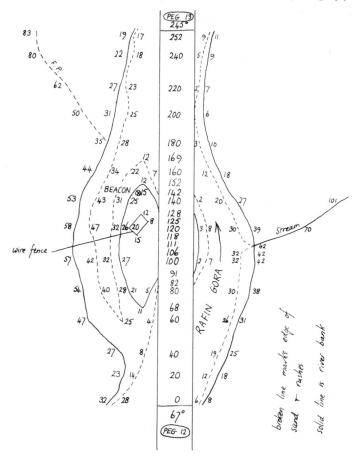

Fig. 53 An example of a field book entry for part of a compass traverse along a shallow river. Entries are made from bottom to top of the page. The centre column represents the measuring tape, and the entries therein are station names, forward and back bearings, and distances along the traverse line. Note that every 20-metre length has been entered to ensure no errors of omission. No Abney clinometer readings have been taken, firstly because the survey line is only very gently inclined, and secondly because corrections to measured lengths would be too small to be plottable on the scale of the finished map.

Some authorities insist that every offset figure should be entered on the chain side of the detail to which it refers. This is an excellent practice if it does not confuse. In the illustration all the figures for offsets to the river bank are on the outer side from the chain in the interest of legibility. Commonsense will usually determine the location of figures, but the result should be clear to other users as well as the booker.

traverse, but it will displace by an equal amount all subsequent points in the traverse. The length of each leg, corrected to the horizontal by the use of Abney clinometer readings only if the map scale justifies such refinement, is scaled on the plotted ray. Detail is not plotted at this stage.

It is unlikely that the traverse when drawn will end at the known position of its terminal point, and the plotted position of this point will be some distance away, the distance being the misclosure of the traverse. As compass traverses are among the least accurate forms of survey, the misclosure may be as much as 0.02 of the total distance traversed, i.e. 20 metres per kilometre. If the misclosure exceeds 2%, the compass should be tested for accuracy over a line of known bearing. If the compass reads consistently high or low, the traverse plot may be turned through a corresponding angle, which will reduce the apparent misclosure. However, if the misclosure still exceeds the 2% figure, and no error of plotting can be found, the traverse should be rejected. An acceptable misclosure is adjusted by Bowditch's Rule, already used mathematically for the adjustment of theodolite traverses (see p. 63), but applied graphically in the correction of compass traverses. Bowditch's Rule assumes that the error has been accumulated evenly throughout the traverse, and amount of adjustment is therefore proportional to the distance travelled. Thus in fig. 54 the total misclosure is the distance $Q'Q$: the end of the traverse has to be moved from Q' to the true position Q. Each of the other traverse points must be moved parallel to

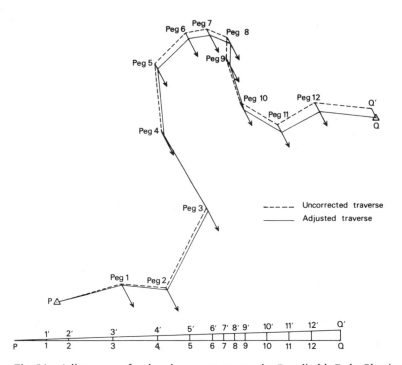

Fig. 54 Adjustment of a closed compass traverse by Bowditch's Rule. Plotting of bearings and distances consecutively from P leads to location of Q' some little distance from the correct position Q. Provided that $Q'Q$ is less than $\frac{1}{50}$ of the traverse length, the traverse can be adjusted. All points move parallel to $Q'Q$ and in the same direction, the amount of each shift being taken from the triangle in the illustration. Lengths $P1, 12, 23 \ldots 12Q$ are proportional to the traverse lengths between the corresponding points, and the perpendicular QQ' is made equal to the misclosure. The other perpendiculars are the true shifts to be made for the location of the respective pegs.

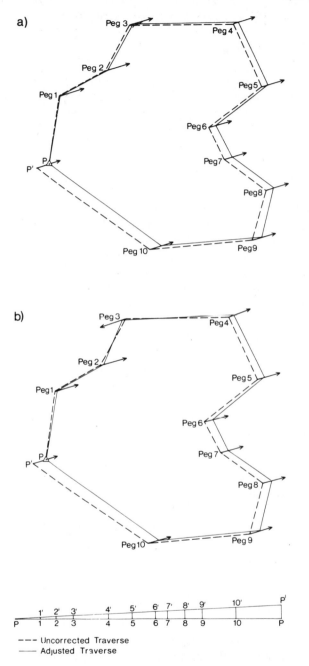

Fig. 55   (a) The correct adjustment of a traverse, all points moving in the same direction. (b) An incorrect attempt at adjustment, as the first plotted location of Peg 3 has been moved wrongly. The crossing of the adjusted and unadjusted lines between Pegs 2 and 3, and between Pegs 3 and 4, is a certain indication of error.

Q'Q and in the same direction: the amount of each movement is

$$\frac{\text{traverse distance to peg}}{\text{total traverse length}} \times \text{total misclosure.}$$

This is obtained graphically. A horizontal line PQ is drawn representing the total traverse length, and at Q a perpendicular QQ' equal to the misclosure. PQ' is drawn. At each of the intermediate pegs 1, 2, 3, 4, etc., a perpendicular is drawn to meet PQ' at $1'$, $2'$, $3'$, $4'$, etc. The lengths $11'$, $22'$, $33'$, $44'$, etc., are the adjustments to be made at the corresponding traverse points. In constructing the line $P1234 \ldots Q$ it is not necessary to draw PQ the actual length of the traverse if this is too long for convenient drawing. PQ can be drawn to scale, e.g. one-fifth of the plotted length, and $P1$, 12, 23, etc., are also scaled along the line at one-fifth actual length. The perpendiculars, however, are kept at true length in order to allow direct measurement with dividers or scales for the transference to the plotted traverse.

Each correction having been applied, the traverse is replotted between the adjusted station positions, and when this has been completed the adjusted traverse line is transferred to the main map. Offset detail can now be added to the traverse lines, plotted directly from the field book entries. As the adjustment is mainly one of angular correction, the lengths of traverse legs should not be appreciably altered, so that offset plotting should present no difficulties. If it is found that the length of an adjusted leg differs widely from the measured length entered in the field book, there has been an error in the plotting: either the length was incorrectly scaled in drawing the traverse before adjustment, or the amount of station adjustment has been incorrectly measured at one point, or the adjustment at one point has been made in the opposite direction to that of the misclosure (in this case two adjacent traverse legs will reveal errors of length, and the error should be readily visible by inspection; fig. 55 shows correct and incorrect traverse adjustment, and it can be seen at once that the incorrect movement of station 3 leads to a double crossing-over of preliminary and adjusted traverse plots).

For small-scale work, compass traverses can be very useful even when combined with unorthodox methods of measurement. Pacing has already been mentioned. A bicycle wheel of known circumference will give lengths along a traverse within accuracy limits of small-scale work: a piece of string tied round the tyre allows easy counting of the number of wheel revolutions in each line even if the machine is ridden and not pushed. In a reconnaissance traverse forward bearings are taken to the next point likely to be useful, and back bearings are taken to the last point occupied. Offsets are by pacing. The work can be done by a single individual, and no poles are used. The observer follows a straight line towards the point to which he has taken the forward bearing. He can mark each station, for back bearing purposes, with a small piece of paper on a twig if he feels in doubt about identification of the position.

Heights in conjunction with compass traverses

The strip map produced by a compass traverse can be greatly enhanced by the extensive use of an Abney clinometer. Generally speaking, compass traverses are of such low accuracy that corrections of measurements for slope are a waste of time, the corrections being too small to be plottable. However, slope readings can be used for the calculation of heights round the traverse and for the sketching of formlines on either side, and it is recommended that the Abney clinometer should be used for this purpose.

Slope readings should be taken:

(a) directly from each traverse station to the next;
(b) over each chain or tape length, or to each break of slope along the traverse;
(c) at regular intervals, or at breaks of slope, at right angles to the line of the traverse, with some indication of the distance for which the slope reading is estimated to be valid.

Figure 56 represents a simplified section of a traverse, showing traverse stations A, B and C. θ is the angle XAB, the angle of elevation of B as observed directly from A. ϕ is the corresponding angle of depression of C from B. The slope AB is not uniform, however, but follows the profile $ApqrB$, and the distances chained will be along this profile. If the example applied to a theodolite traverse, a slope reading would be taken for each of the sections Ap, pq, qr and rB, the length of each would be adjusted to the horizontal, and the sum of these values would give the true horizontal distance AX. The difference in height between A and B, which is BX, would equal $AX \tan \theta$, θ being measured at both A and B with the vertical circle of the theodolite.

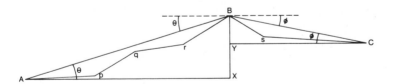

Fig. 56   For rough estimation of altitude differences BX and BY, and to obtain altitudes along the traverse ABC, these differences can be regarded as AB . $\sin \theta$ and BC . $\sin \phi$, AB and BC being the respective totals of measured distances $ApqrB$ and BsC.
More accurately, one can calculate altitudes along the traverse by using Ap . $\sin$ (slope angle Ap), pq . $\sin$ (slope angle pq), etc., or as AX . $\tan \theta$ and YC . $\tan \phi$ after correction of measured distances to the horizontal. However, Abney clinometer readings over all measured lengths are often omitted with compass traverses unless the slopes exceed $5°$.

With a compass traverse, Abney clinometer readings are still taken for each of the sections Ap, pq, qr and rB, as well as for θ at A and at B, but unless slopes are steep the horizontal corrections are not made. In calculating traverse heights, it is customary with a theodolite traverse to multiply the horizontal length of the leg by the tangent of the angle of slope from station to station; because one already has in the calculation the logarithm of the true horizontal distance this is the most convenient method. With the compass traverse one is normally using uncorrected lengths, and these are multiplied by the sine of the slope angle to obtain the difference in height between stations. It is obviously important to note whether the angle is an elevation or a depression. When the heights have been calculated for the traverse, the computed altitude of the known point on which the traverse closes is probably not identical with the actual altitude already recorded. The misclosure in height is adjusted by Bowditch's rule as described on page 100, and the altitude of each traverse station is found.

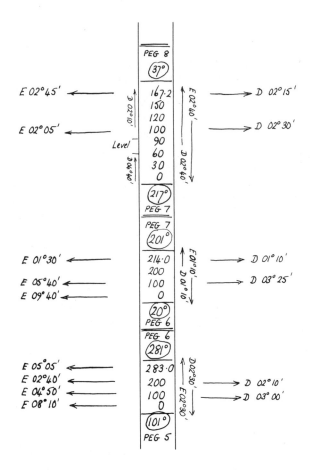

Fig. 57   Each traverse leg should have a separate page in the field book, but three legs are included together here for the convenience of the reader.
    Abney clinometer readings are taken
(a) directly from station to station;
(b) between breaks of slope where the ground is uneven, as between Peg 7 and Peg 8;
(c) at right angles to the traverse line at every 100 m.
    The text describes the use of these slope readings to plot the formlines seen in fig. 59.

Reference is now made to the Abney clinometer readings other than the direct sightings from station to station. Firstly, those taken along the traverse line are used with the relevant lengths, e.g. the field book entries of fig. 57 can be used to calculate intermediate heights between the traverse stations Peg 5 and Peg 8. The slope from Peg 5 to Peg 6 was uniform. The height of Peg 5 after calculation round the traverse was 219.62 m, and the height of Peg 6 was 232.04 m. The distance Peg 5 to Peg 6 was 283 m, and over this distance there is an altitude rise of 12.42 m. The 220-metre contour is 0.38 m above Peg 5, and this rise of 0.38 m will require a distance of 8.7 m; the 220-metre contour therefore crosses the traverse line at 8.7 m from Peg 5. A rise of another 10.00 m requires a distance of 228.9 m, so the 230-metre contour is crossed at a distance from Peg 5 of 8.7 + 228.9 = 237.6 m. The slope from Peg 6 to Peg 7 is also uniform, and is 214 m in length. Peg 6 is at an altitude of 232.04 m and Peg 7 at 227.54, a fall of 4.50 m. If the total fall is 4.50 in 214 m, a fall of 2.04 (to reach the 230-metre contour) will require a distance from Peg 6 of $\frac{204}{450}$ x 214 = 97.0 m.

From Peg 7 to Peg 8 the slope was not uniform, as seen from fig. 57. The first 30 m, with a depression of 04° 40', will show a fall of 2.43 m; the second 30 m, with the same slope, will show a fall of another 2.43 m; the third 30 m length was level, with no change in altitude; the fourth 30 m dropped at 02° 10', giving a fall of 1.13 m; the fifth 30 m shows a similar fall; the final 17.22 m will give a fall of 0.65 m. The overall decline in altitude should therefore appear to be 7.77 m. The overall angle of D. 02° 40' for the 167.22 m from Peg 7 to Peg 8 gave a height difference of −7.78 m, but adjustment of heights for misclosure round the traverse resulted in altitude values of 227.54 m for Peg 7 and 219.68 m for Peg 8. There is thus a fall of 7.86 m from Peg 7 to Peg 8, and the heights calculated from the chain lengths and slopes can be adjusted between the two stations, again by the use of proportional parts:

Station	Chained Distance	Abney Reading	Height Difference	Height from Traverse	Calculated Heights	Correction	Adjusted Heights
Peg 8	167.22	D. 02° 10'	−0.65	219.68	219.77	−0.09	219.68
	150.00	D. 02° 10'	−1.13		220.42	−0.08	220.34
	120.00	D. 02° 10'	−1.13		221.55	−0.06	221.49
	90.00	Level	nil		222.68	−0.05	222.63
	60.00	D. 04° 40'	−2.43		222.68	−0.03	222.65
	30.00	D. 04° 40'	−2.43		225.11	−0.01	225.10
Peg 7	0.00			227.54			

From these figures it can be found that the 220-metre contour was crossed at a distance of 159.0 m from Peg 7 towards Peg 8. (The ground is at 220.34

at 150 m; it falls 0.65 m in the next 17.22 m, and so will fall 0.34 m in 9.0 m.)

Formlines may now be plotted on either side of the traverse. At Peg 5 there is a slope to the left of the traverse line, and at right angles to it, of E. 08° 10'. Peg 5 is at 219.62 m. The tangent of 08° 10' is 0.1435. To

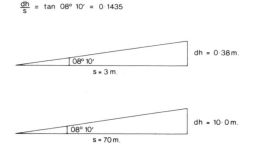

Fig. 58   Calculation of horizontal distances when the slope is 08° 10', to find distance from a point at 219.62 m to the contour at 220.00 m, and the additional distance to the 230.00 m contour.

reach the 220-metre formline there is a rise of 0.38 m. As shown in fig. 58, $\frac{0.38}{s} = 0.1435$, or $s = \frac{0.38}{0.1435} = 3$ m. For a further rise of 10 m the distance will increase by $\frac{10.00}{0.1435} = 70$ m. The 220-metre formline is therefore at a distance of 3 m, and the 230-metre formline at a distance of 73 m. At 100 m from Peg 5 the left-hand slope was E. 04° 50', with a natural tangent of 0.0846. At this point the height on the traverse line is 224.06 (calculated proportionally between Peg 5 and Peg 6). The rise to the 230-metre formline is therefore 5.94 m, and the distance is $\frac{5.94}{0.0846} = 70$ m. On the right-hand side of the line the slope is D. 03° 00', with a natural tangent of 0.0524. The drop to the 220-metre formline is 4.06, and the distance is $\frac{4.06}{0.0524} = 78$ m.

At 200 metres from Peg 5 towards Peg 6 the height of the traverse line can be calculated as 228.12 m. The left-hand slope is E. 02° 40', tangent 0.0466: the distance to the next formline is $\frac{230 - 228.12}{0.0466} = \frac{1.88}{0.0466} = 40$ m. If the slope angles shown in the field book (fig. 57) are used in this way, the formlines of fig. 59 will be obtained.

This is a very simple way of adding approximate formlines. The taking of the extra Abney readings adds little to the time of the survey, and although the calculations take time, the value of the traverse is considerably increased.

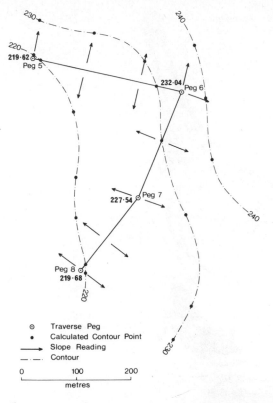

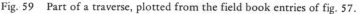

Fig. 59   Part of a traverse, plotted from the field book entries of fig. 57.

The Brunton compass and inclinometer (or pocket transit)

The Brunton compass is manufactured in the United States of America, and combines many of the features of the Prismatic Compass and of the Abney Clinometer. It is a very ingenious instrument which, with its collapsible tripod, is easily carried in a small sling case.

The instrument itself is approximately three inches square. It can be fastened to a bracket on the tripod head by means of a clamp screw (see fig. 60a) which grips the side of the instrument case.

Use as a compass

After attachment to the tripod, the lid of the case is folded back. The inside of the lid is mirrored, apart from a small oval near the hinge, which is open. A centre line crosses the entire lid as shown. A small vane folds over the edge of the lid to form the backsight. A tapered bracket hinges in the opposite direction to the lid, to form the foresight arm: the tip of this can be turned up as the foresight.

The main instrument case contains a compass needle mounted on a central pivot: a small stud, which is depressed when the lid is closed, actuates

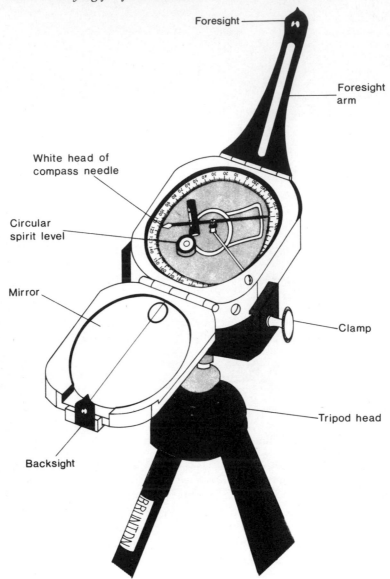

Foresight

Foresight arm

White head of compass needle

Circular spirit level

Mirror

Clamp

Tripod head

Backsight

BRUNTON

Fig. 60a   The Brunton Transit used as a compass. As the angle of sight in this case is inclined upwards, the foresight arm is partially elevated. The horizontal scale is actually graduated in whole degrees, but the diagram shows only 5° intervals.

a lever which lifts the needle from its pivot, as with the Prismatic Compass. The needle is painted black, but with a white head at the North end. This facilitates reading against a circular scale graduated in degrees. A circular spirit level allows for movement of the instrument on a ball and socket joint

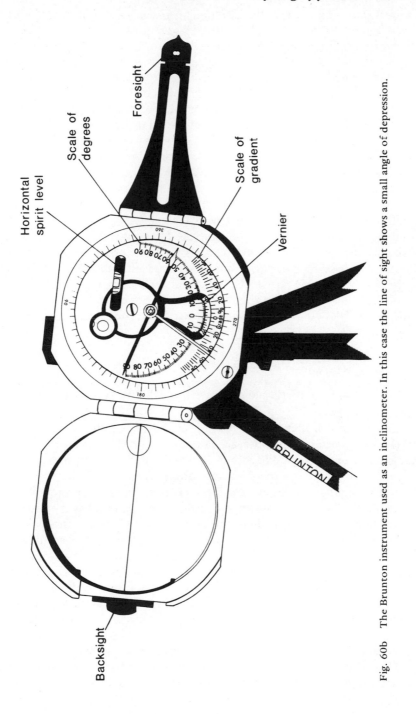

Fig. 60b   The Brunton instrument used as an inclinometer. In this case the line of sight shows a small angle of depression.

on the tripod head to bring the case to a horizontal position ready for use.

The points of the backsight and foresight are lined up with the target, and the bearing then read from the needle head. The 0° mark lies on the line between the sights, and the fixed circular scale is graduated anti-clockwise to 360°. As the needle when free to move lies on the magnetic meridian, its point therefore measures the magnetic bearing clockwise from North to the target. The foresight arm can be raised or lowered to allow sights to be made with direct alignment even if the angle of elevation reaches 30°. If the angle of elevation is greater, the backsight can be turned vertically, and sighting made through the oval hole, aligning the marker line with the tip of the foresight, for angles up to about 70° of elevation. Alternatively, the mirror can be set at an appropriate angle and the target image is viewed from the foresight, which is lined up with the reflected image and the centre line. The needle head still gives a correct magnetic bearing.

For bearings over lines of depression, the foresight is put in the horizontal position and the backsight raised. This allows direct observation for angles to about 40°. For steeper angles, it is necessary to view from elevated foresight to the oval hole in the base of the lid, which allows sightings as steep as 70° below the horizontal, but in this case the bearing read must be altered by 180°, as one is of course obtaining the back bearing from the needle when using the instrument in reversed direction.

Use as an inclinometer

The tripod head can be turned sideways through 90° to bring the instrument case into a vertical position. With the lid open, and the foresight extended the instrument will then appear as in fig. 60b. The instrument is so designed that it can be turned only when the horizontal spirit level is at the top (the clamp screw on the tripod head, not shown in the figure, impedes any attempt to turn the instrument into an inverted position).

The backsight and foresight are aligned with the target — preferably with a point on the target staff at the same height above ground as the instrument. A small lever on the underside of the case (i.e. on the back in this position) rotates the bracket on which are mounted both the horizontal spirit level and the vernier arm. The bracket is turned until the bubble is central in its tube, and the zero of the vernier scale indicates the angle of inclination. Whole degrees are read from the main scale on the inside of the instrument case, and the vernier gives the degree fraction to the nearest 10 minutes. Alternatively the zero mark may be read against the scale of percentage gradient.

The lid of the case has on it a table of natural sines (to three decimal places) for whole degrees from 0° to 45°. This allows immediate calculation in the field of height differences between traverse points, using measured distances and slope angles obtained.

Use of the aneroid barometer

Another instrument for height determination is the aneroid barometer, and this can be used in conjunction with a compass traverse, although it may be used independently. The aneroid is a very sensitive instrument, which responds to very small changes in atmospheric pressure and can measure differences in altitude of less than half a metre if two points are visited in quick succession in a period of atmospheric stability. Any reading of the aneroid includes two errors:

(a) instrumental error, giving a figure above or below the true value;
(b) error due to variation in atmospheric pressure.

The instrumental error is a constant which could be applied to every observation taken, if its magnitude were known. The second error is variable due to the diurnal pressure wave, so that even if the aneroid is left undisturbed throughout the day its readings may vary by as much as 60 or 70 metres. The constant instrumental error is therefore impossible of measurement: if the aneroid is held at a Bench Mark of known height the position of the needle may be adjusted to give the correct reading, but within a few minutes it will be recording above or below this figure. It may be assumed that the instrumental error has been eliminated, but the diurnal variation is then likely to move between positive and negative values, increasing the

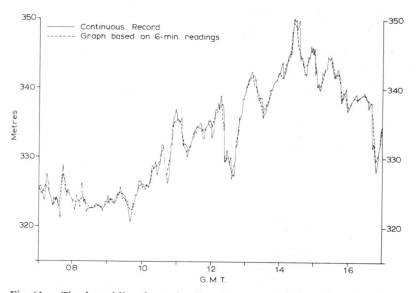

Fig. 61a   The dotted line shows the diurnal wave as plotted from readings taken with an aneroid at regular intervals of 6 minutes. The continuous line shows the record that would have been obtained with a continuous trace from a sensitive barograph. The two lines agree closely over most of the day, but with some discrepancies of as much as 3 m (07.38 hours) or 4.5 m (14.35 hours and 16.37 hours).

likelihood of confusion and of error in calculation. Moreover, if working at low altitudes, apparent elimination of the instrumental error may later lead to the needle falling below zero, where many instruments have no graduations, so that no reading is possible.

It is more convenient to combine the two errors into a single variable, and to set the instrument to read appreciably too high, so that the needle always stays well above zero. This has the additional advantage that it will be obvious at any time whether an altitude reading made with the aneroid is adjusted or unadjusted for error. The observer may have used the instrument in an area of between 100 and 200 m altitude above mean sea level, with the aneroid reading between 450 and 550 m. A value of 527 m, for example, is then obviously a direct instrumental reading that has not been adjusted for the total error.

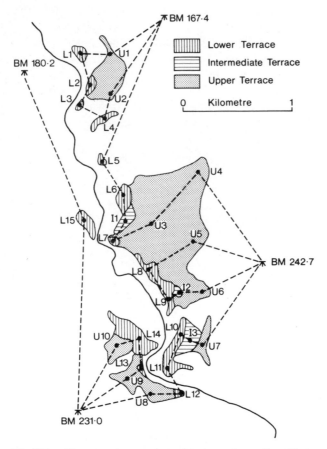

Fig. 61b   The map shows terrace fragments in a valley. The route taken by the surveyor with the aneroid is shown by the broken line, and all terrace points at which observations were made are shown by dark spots. It can be seen that the route made repeated calls at Bench Marks for check readings against known altitudes. See fig. 62 for the corresponding field book entries.

Survey: R. Ciwo Terraces
Date: 12. 12. 72
Observer: T. Thomas

Aneroid
Traverse

STATION	INSTRUMENT READING	TIME	LEVELLED HEIGHT	ERROR or CORRECTION	CALCULATED HEIGHT	REMARKS
BM 167·4	323·8	08. 00	167·4	156·4		
U 1	328·3	08. 11				
L 1	308·1	08. 16				
L 2	311·1	08. 23				
L 3	311·6	08. 27				
L 4	317·2	08. 33				
U 2	331·7	08. 38				
BM 167·4	323·3	08. 54	167·4	155·9		
L 5	324·5	09. 13				
L 6	331·2	09. 20				
I 1	348·9	09. 24				
L 7	336·5	09. 28				
U 3	357·5	09. 33				
U 4	362·3	09. 41				
BM 242·7	400·3	09. 51	242·7	157·6		
U 5	376·1	10. 00				
L 8	344·0	10. 11				
L 9	353·2	10. 20				
I 2	367·9	10. 24				
U 6	376·7	10. 32				
BM 242·7	403·0	10. 41	242·7	160·3		
U 7	382·9	10. 52				
I 3	380·4	10. 55				
L 10	365·5	10. 58				Strong gusts of wind
L 11	368·8	11. 06				—do—
L 12	370·2	11. 15				
U 8	387·6	11. 18				
BM 231·0	395·8	11. 25	231·0	164·8		
U 9	387·3	11. 34				
L 13	367·6	11. 39				
L 14	361·4	11. 46				
U 10	382·8	11. 52				
BM 231·0	399·9	12. 06	231·0	168·9		
L 15	339·7	12.33				Sharp shower
BM 180·2	352·8	13. 01	180·2	172·6		

Fig. 62   The field book for the aneroid traverse plotted in fig. 61b. Each check observation at a Bench Mark enables the total error of the instrumental reading to be calculated: this includes both the instrumental error and the diurnal pressure wave error.

If the observer has two aneroids, the calculation of error for correction is relatively simple provided that he has an assistant available to remain at his base throughout the day and to take an aneroid reading every five minutes. These observations can later be plotted against time, to provide a picture of the pressure variation throughout the day. If we suppose that the base station is known to have an altitude of 167.43 m, and the recordings taken throughout the day give the values plotted in fig. 60, it is possible to determine the total combined error for any time of day, e.g. at 09.30 hours the aneroid read 324.5 m, and at 15.42 hours it read 342.2 m. The total errors at the two times were respectively 157.1 and 174.8 m. The second aneroid was also read at base before setting out, giving 281.3 m at 09.30 hours, i.e. with a total combined error of 113.9 m. Clearly, the second instrument has an instrumental constant error which is 43.2 m less than that of the first instrument. If the second instrument has been read at the base at 15.42 hours, it would again have read 43.2 m less than the first instrument, i.e. 342.2 − 43.2 = 299.0 m. However, the second instrument was read at the level of a stream confluence some distance away at 15.42 hours, giving 236.7 m. The height of the confluence is easily found: at 15.42 hours the total error was 174.8 m on the first instrument, or 174.8 − 43.2 = 131.6 m on the second instrument. The correct height for the confluence was therefore 236.7 − 131.6 = 105.1 m. Similarly, the height of any other point observed with the second instrument can be determined: the correction will be, at any particular time,

(1st instrument correction *for that time*) − (43.2)

Usually the observer has only one aneroid, and therefore cannot establish a control wave so easily. Instead he has to combine observations made at points for which he requires altitude values with those made at known altitudes which he uses as a check on the diurnal wave. He must start and finish at points of known altitude, and must check his instrument against a known altitude at least once per hour and preferably more often. A geographer obtaining altitudes of terrace fragments in a valley such as that illustrated in fig. 61b will produce a field book as in fig. 62. This shows the

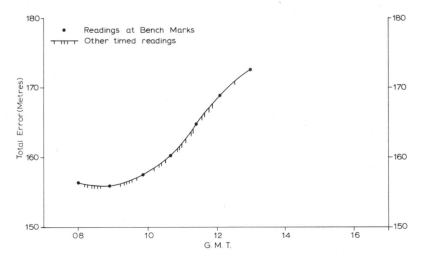

Fig. 63a   The Bench Mark readings entered in the field book of fig. 62 are plotted, and a curve drawn to join them together as an approximate diurnal wave. The times of the other readings are shown also.

The graph shows Total Error, as that is what the surveyor wishes to know in respect of every reading taken, and Total Error can be found irrespective of the Bench Mark at which a check is made.

entries for all the terrace points marked with a black spot in fig. 61b and for the readings at Bench Marks. The diurnal wave is plotted from the readings made at the Bench Marks, giving an approximate wave as in fig. 63a. The altitudes of the intermediate points can be obtained by scaling errors from this graph for the appropriate times and applying them as corrections to the observed values, as in fig. 63b. Note that one does not plot *altitudes* against time, because the periodic visits to Bench Marks may use different Bench Marks as the day's work proceeds. One plots *total error* against time, as the total error can be calculated at any known altitude by subtracting the true value from the instrumental reading.

Survey: R. Ciwo terraces
Date: 12.12.72
Observer: T. Thomas

Aneroid
Traverse

STATION	INSTRUMENT READING	TIME	LEVELLED HEIGHT	ERROR or CORRECTION	CALCULATED HEIGHT	REMARKS
BM 167.4	323.8	08.00	167.4	156.4		
U 1	328.3	08.11		156.2	172.1	
L 1	308.1	08.16		156.1	152.0	
L 2	311.1	08.23		156.0	155.1	
L 3	311.6	08.27		155.9	155.7	
L 4	317.2	08.33		155.8	161.4	
U2	331.7	08.38		155.8	175.9	
BM 167.4	323.3	08.54	167.4	155.9		
L 5	324.5	09.13		156.3	168.2	
L 6	331.2	09.20		156.4	174.8	
I 1	348.9	09.24		156.6	192.3	
L 7	336.5	09.28		156.7	179.8	
U 3	357.5	09.33		156.8	200.7	
U 4	362.3	09.41		157.2	205.1	
BM 242.7	400.3	09.51	242.7	157.6		
U 5	376.1	10.00		157.9	218.2	
L 8	344.0	10.11		158.4	185.6	
L 9	353.2	10.20		158.9	194.3	
I 2	367.9	10.24		159.2	207.8	
U 6	376.7	10.32		159.7	217.0	
BM 242.7	403.0	10.41	242.7	160.3		
U 7	382.9	10.52		161.3	221.6	
I 3	380.4	10.55		161.6	218.8	
L 10	365.5	10.58		161.9	203.6	Strong gusts of wind
L 11	368.8	11.06		162.7	206.1	—do—
L 12	370.2	11.15		163.8	206.4	
U 8	387.6	11.18		164.1	223.5	
BM 231.0	395.8	11.25	231.0	164.8		
U 9	387.3	11.34		165.8	221.5	
L 13	367.6	11.39		166.2	201.4	
L 14	361.4	11.46		166.9	194.5	
U 10	382.8	11.52		167.5	215.3	
BM 231.0	399.9	12.06	231.0	168.9		
L 15	339.7	12.33		171.0	168.7	Sharp shower
BM 180.2	352.8	13.01	180.2	172.6		

Fig. 63b   The field book shown in fig. 62 can now be completed. The Total Error of each reading is taken from the graph of fig. 63a and is entered as a correction to be subtracted from the Instrument Reading to give the Calculated Height.

In the field, care must be taken to treat the instrument gently under all circumstances, as even a moderate jarring will throw its compensating system into disarray and will make the instrument unusable. The aneroid should always be held alongside the Bench Mark, at the same height as the official mark, and it should be allowed a brief interval, perhaps 30 seconds, to overcome the hysteresis effect which delays the response to pressure changes. The aneroid should *not* be given robust taps in an effort to accelerate response.

The single-instrument method described here is, of course, less accurate than the two-instrument method with a control wave based on 5-minute readings. A graph based on intervals of 45 to 60 minutes misses many slight oscillations of pressure, and is thus liable to errors of some metres. For this reason, 7 to 8 m is generally considered to be the attainable limit of accuracy. Figure 63c indicates how "smoothing" of the graph by less frequent check readings inevitably introduces errors; also minor variations of topography can affect readings, particularly under conditions of strong wind, so that an instrument which would show a total error of x metres if used at one point may show an error of y metres when used at another. If an obvious cyclonic front passes over the work area, readings since the last Bench Mark check should be abandoned, a new check made and the points revisited, as a sharp change in pressure and therefore in altitude reading is highly probable. In the two-aneroid method, the distance between the two instruments should not become too great, as the method assumes identity of atmospheric conditions for the two instrument locations. In the British Isles it is unlikely that one day will reproduce the diurnal wave of

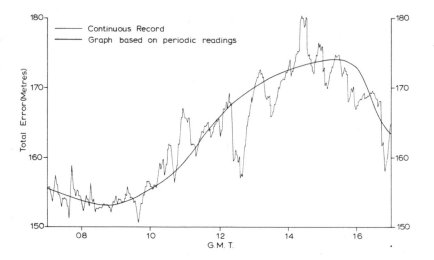

Fig. 63c   The graphs of figs. 61a and 63a combined to show the errors that can arise with the use of a smoothed curve based on infrequent checks at Bench Marks instead of a continuous check from a barograph. Errors can be as much as 11 m (12.36 hours).

the preceding day, but in some climatic zones, at least at some seasons, the diurnal wave will have almost constant form from day to day.

Aneroid observations for altitude can be usefully combined with a compass traverse. If a control wave can be established from Bench Marks or previously heighted points, it may be easier to calculate the aneroid observations than to calculate traverse heights from Abney clinometer observations; the former need correction only for the total error plotted on the control wave, whereas the latter require long calculations and adjustment of any misclosure. However, aneroid readings cannot be regarded as very reliable except under conditions of marked atmospheric stability, and geographers will generally use them only for work of a rapid or reconnaissance nature. The aneroid is more useful for determing differences in height than for determining absolute heights above mean sea level, and can be excellent for mapping erosion surfaces on 1:250,000 or 1:500,000 scales, for example, when observations at 1-kilometre intervals, taken in a moving vehicle, can be very useful in determining surface changes, particularly in country without large-scale contoured maps and with a heavy vegetation cover inhibiting views of any great extent.

Subtense measurement of distance   One optical method of distance measurement has still to be mentioned, namely the use of a subtense bar. This is a horizontal bar with precise measurements of length marked upon it, set on a tripod. It could, for example, be centred over Peg Q and set at right angles to the line Peg P–Peg Q by means of a fitted sight. Assume the length AB on the bar to be 2 m, which is common. A theodolite at Peg P then makes a single measurement of the angle $A\hat{P}B$, which could be 02° 01′ 10″. The observer then makes a multiple measurement of the angle (see p. 24) and obtains an angle of 60° 33′ 10″ for a multiple of 30, so that the angle is redefined as 02° 01′ 06″.33. As Q is midway between A and B, AQ = 1.0 m, the angle $A\hat{P}Q$ = 0.1° 00′ 33″.17, and the angle $A\hat{Q}P$ = 90°. PQ can be calculated, being $AQ \cot A\hat{P}Q$, in this case 56.77 m. No slope correction is required. In place of a subtense bar, a theodolite at Q could set out at right angles to PQ a line AB measured with a tape.

Professional surveyors make considerable use of subtense measurement of distance. Non-specialists may experience some difficulty in avoiding errors due to inaccuracy in the length of AB, of the horizontality of AB, of the angle $A\hat{P}Q$ and of the 90° at Q.

4 Instruments

General principles

Every field scientist should have some knowledge of the design of the instruments he uses, in relation to the purpose for which they were constructed. In some of his techniques he will be making use of only part of the instrumental capability, or will be working to a standard of accuracy which is below the highest standard which the instrument can provide. In other cases he will wish to take observations above the direct instrumental capability, as when he uses a theodolite graduated to 20″ of arc for the measurement of small angles to seconds and tenths. Frequently he will need to employ instruments of different make from those on which he was trained. In all circumstances he can achieve his ends with minimum effort and maximum efficiency if he understands the basic principles of instrument design, knows what controls to expect and what adjustments to make, so that not only can he operate an instrument of unfamiliar detail, but he can test and adjust it before and after use in order to check his observations.

It cannot be stressed too strongly that no instrument will give good results unless it is given due care and attention. Every piece of equipment must be put away after use in a clean and dry condition, ready for immediate use when required. Instruments should always be checked before use to ensure that they are in adjustment, and should be similarly checked before return to store: if employed over a fair period of time, they should be checked at intervals and adjusted if necessary.

The level

Levels are obtainable in a variety of forms, but all embody one essential principle: the instrument consists of a telescope which can be brought accurately to the horizontal position, the line of sight being defined by a horizontal crosswire when viewed through the telescope. The design must therefore include some means of checking that the line of sight is truly horizontal.

Precise instruments have as a base a levelling head with three footscrews similar to a theodolite, and used in similar fashion (see p. 134). The observer will most probably be using a Quickset level, in which the levelling head is replaced by a ball-and-socket joint. This is locked by a milled ring (see fig. 64). When the ring is slack, the whole telescope system may be moved on the ball, and it is moved until a small circular spirit level to one side of the telescope shows the bubble in the centre of the small circular graticule. With this bubble in centre the ring is tightened and the instrument locked on the ball socket. The telescope is then free to turn in any direction, until it is fixed by a small clamp screw which fastens the telescope in relation to the vertical axis. Some manufacturers fit a small horizontal circle, graduated in degrees, with an index mark, and this can be used to set out angles to the nearest half-degree: however, this is very rarely used except for rough alignment by engineers (for whom the instrument is principally designed), as a theodolite will normally be employed if horizontal angles are required.

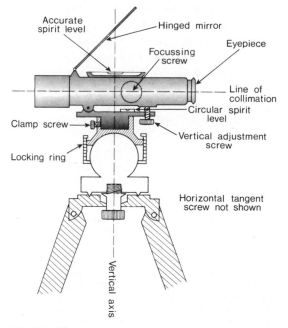

Fig. 64   The component parts of the quickset level.

With the instrument approximately levelled, attention is turned to the telescope system. Looking through the telescope, the crosswires are focussed by turning the eyepiece until they are clear: this is best done with the crosswires against a plain surface, which can be a hand or a piece of paper held in front of the telescope. With the crosswires clear, turn the telescope until it is approximately on line with the graduated staff, and clamp with the staff in the field of view. Focus the staff with the focussing screw, which is usually on the side of the telescope but is sometimes a milled ring forming part of the telescope case. When in focus, move the eye slightly up and down while continuing to look through the telescope. The horizontal crosswire should remain unmoving on the staff graduations. If it moves over the staff, there is a slight parallax error, which can be eliminated by slight adjustment of the focus (or if this does not eliminate it, by slight adjustment of the eyepiece). When there is no parallax, the vertical crosswire is brought laterally to the centre of the staff by use of the slow-motion tangent screw: this can be identified by its horizontal attitude, set to push the telescope sideways. It will move the telescope only when the clamp screw is fastened. If the tangent screw reaches the end of its run, the clamp screw is loosened and the tangent screw turned back to the middle of its threaded length, when the clamp can be refastened.

Before a reading is attempted, the line of sight must be truly horizontal. Mounted on the telescope is a sensitive spirit level, which can usually be observed from the eyepiece end of the telescope by means of an inclined mirror above the spirit level. The bubble is brought to centre by means of a vertical slow-motion screw which raises or lowers one end of the telescope: it can be identified by its vertical attitude. With the bubble centred, the reading on the staff made by the horizontal crosswire is then taken, followed (as a check, see p. 45) by the readings of the upper and lower stadia wires, these being the two short horizontal lines shown in fig. 65. The accuracy of the readings depends on the distance from telescope to staff and the power of the telescope. On short lines readings can be taken to the nearest millimetre, by estimation; on longer lines one uses the 5 mm graduations. If

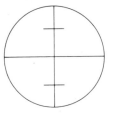

Fig. 65   The graticule seen through the telescope. The vertical line is single. There are three horizontal lines, one being the main crosswire to which readings are taken, and two short stadia lines, one above and one below. Readings to the stadia lines are used firstly as a check on the crosswire reading (because Upper Stadia–Crosswire should equal Crosswire–Lower Stadia) and secondly for the tacheometric measurement of distance [(Upper Stadia–Lower Stadia) multiplied by 100 equals the distance from the telescope to the graduated staff].

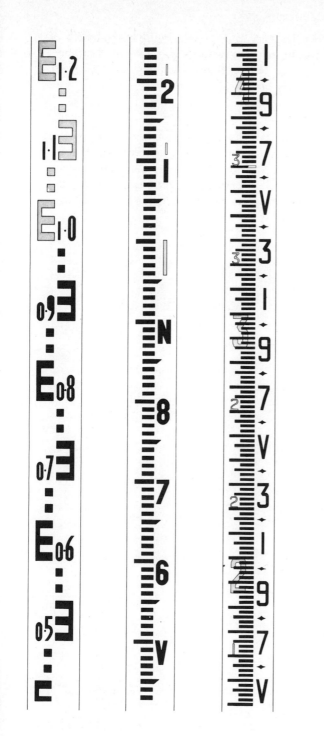

Fig. 66   Three different markings of levelling staves.

In the centre is a standard metric pattern, with graduations to 0.005 of a metre. Lettering is in black except for metre readings, which are shown by red figures at the exact metres and by red dots or miniature figures above the black numbers used to indicate tenths of metres. Note that the number six is shown by 6, which can be mistaken for nine when viewed in the inverted position as seen through the telescope. Nine is shown by the letter N.

The staff on the left is also metric, but with hundredths shown in an oddly grouped pattern. Alternate metres are shown in red and black, but all figures for metres and tenths are in black.

The staff on the right is graduated in feet and hundredths, with red numbers to show the feet. British practice puts the top of a number alongside the appropriate graduation, e.g. 2.700 feet is shown by the edge of the long graduation corresponding to the top of the horizontal stroke of the figure 7.

using a staff graduated in feet the readings are to the nearest hundredth of a foot. As described on p. 46, the distance from telescope to staff can be measured optically: it equals 100 x (Upper Stadia Reading − Lower Stadia Reading).

Levelling staves

The two types of staff most likely to be used by geographers are the hinged staff and the extendable Sopwith staff. The former is lighter, shorter and easier to carry, but is more liable to shakiness and error, and more susceptible to damage. The Sopwith staff consists of three sections in telescopic form, the total height, if all sections are extended, being 14 ft or 4.27 m.

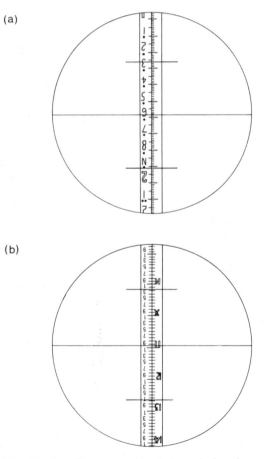

(a)

(b)

Fig. 67　Levelling staves viewed through the telescope, when the image is inverted.
(a) The crosswire on the metric staff reads 1.585 m, and the stadia lines 1.910 and 1.260 m, giving a distance to the staff of 65.0 m.
(b) The staff has a crosswire reading of 10.92 ft, with stadia readings of 12.67 and 9.16 ft, giving a distance to the staff of 351 ft.

It has the advantage of stability and the disadvantage of weight. Both types have graduations on paper strip which can be renewed if scarred or dirty. The patterns of graduation available also vary. Figure 66 illustrates the two types most commonly used in the United Kingdom. The top of a figure indicates the graduation to which the figure refers. The fractions are in black, the whole feet or metres in red: large red figures indicate whole units, small red figures or dot combinations are to assist the observer on short lines when there may not be a whole unit figure in the field of view. Figure 67 shows readings of 1.585 m and 10.92 ft, with stadia readings of 1.910 and 1.260, and of 12.67 and 9.16, respectively.

It is advisable to go into the field with two short lengths of Scotch tape for use in contouring, etc. (see pp. 94 and 95). These can be attached to the side of the staff and transferred to cross particular graduations on the face as required.

The staff should always be held in a vertical position, as any forward or sideways inclination will alter the readings obtained. If the assistant balances the staff between his hands he can maintain the vertical position without difficulty. Some manufacturers fit a small spirit level on the rear of the staff, so that the holder can keep the bubble in a central position to maintain verticality.

Adjustment of the level

The most likely source of error in levelling is one of collimation in altitude; this is when the line of sight from eyepiece to horizontal crosswire is not parallel to the axis of the sensitive spirit level, so that when the bubble is in centre the line of sight is not horizontal. Figure 68 indicates the error in reading that can be made under these circumstances. The importance of trying always to have foresight and back sight lengths equal is stressed on p. 44, and it can be seen from fig. 68 that if these lengths are equal the errors cancel themselves by an equal compensating displacement. However, it is not always possible to balance lengths of sight, particularly when contouring, so that adjustment of the instrument is an important preliminary.

The method of adjustment for collimation is as follows. Set up the level on its tripod at C, exactly midway between A and B. Take readings to A and B. Suppose A gives 1.315 and B gives 1.630. The difference is 0.315 (this is the true difference, errors cancelling out). Move the instrument to D, outside B but close to it, and take readings to A and B. Suppose the reading to A is 1.230 and to B is 1.425: the apparent difference in height is 0.195. Clearly the instrument is out of adjustment. If the B reading of 1.425 is considered to be close to the value that would be obtained with a truly horizontal line of sight, the value at A should be 1.425 − 0.315 = 1.110. Use the vertical tangent screw to bring the crosswire to 1.110, which throws the sensitive bubble off centre. Bring it back to centre by adjusting the fastening and lock nuts at one end or both ends of the spirit level. Check the reading on B. If it has, for example, altered to read 1.420, the value at A must be

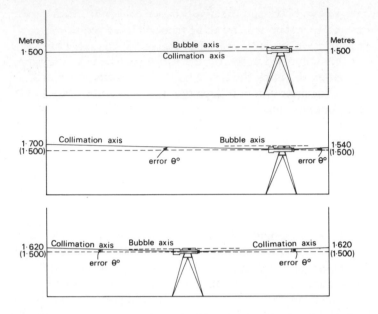

Fig. 68   The effect of an error of collimation in altitude. The top diagram shows the instrument in adjustment, with bubble axis and collimation axis parallel. The lower diagrams show the two axes not parallel. In the middle diagram the tilt on the collimation axis gives readings of 1.700 and 1.540 on the two staves, indicating an apparent difference in altitude of 0.160 m. In the bottom diagram the level is placed centrally between the staves, so that although the collimation axis is tilted, the error is the same on each staff and there is no apparent difference in altitude.

   Although errors cancel out as shown if the level is midway between staff positions, this equality of distance is not always possible, particularly when contouring, so that it is important to keep the level in adjustment.

1.420 − 0.315 = 1.105. Use the vertical tangent screw to obtain this reading; use the adjusting nuts to swing the bubble back to centre.

   The drill for using the level is set out on pp. 125–8. When the full procedure has been followed, the readings taken to the back staff position for crosswire and stadia lines are checked, and the staff is moved to the new forward position as nearly as possible at the same distance from the instrument as before. When readings to this position are completed, the level and tripod are moved to a new point beyond the forward staff.

   At the starting point a height must be taken from a Bench Mark. If this is a brass stud at ground level, the staff can be stood upon it. If the mark is above ground level, the staff is stood alongside it, at station A, for example, and the staff graduation opposite the mark is noted. If the Bench Mark height is 167.308 m above Mean Sea Level, and the staff reading opposite the mark is 0.783 m, station A is at 167.308 − 0.783 = 166.525 m altitude. If the back sight to the staff at A reads 1.310 m, the line of sight is at 166.525 + 1.310 = 167.835 m, and this is the altitude of the telescope. If

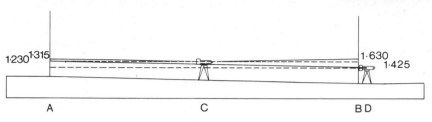

Fig. 69   A method of testing for collimation error. When the level is at *C* it is exactly midway between staff *A* and staff *B*, and although the line of collimation (shown as a full line) is not coincident with the horizontal (shown by the broken line) the errors of reading on *A* and *B* will cancel out as seen in fig. 68. The true altitude of *B* relative to *A* is therefore 1.630 − 1.315 = −0.315 m.

When the level is established at *D* and readings taken to *A* and *B*, the error introduced at *B* is small but that at *A* is large. The difference in readings suggests that *B* is lower than *A* by 0.195 m.

The true difference being −0.315, the reading on *A* should be 1.110, assuming that *B* will read 1.425, the level still being at *D*. Use the vertical tangent screw to bring the crosswire to read 1.110, and then adjust the screws holding the spirit level until the bubble is central.

the reading on the forward staff *B* is 1.485 metres, the base of the staff at *B* is at 167.835 − 1.485 = 166.350 m.

Method of booking

There are two different methods of booking levelling observations. The most suitable method for the non-specialist scientist who wishes not only to find heights of specific fixed points but also the location of fixed heights (in the form of contours) is the height of collimation method. Figure 70 shows a booking form. Each line represents one staff station and the staff readings to it, or one instrument position. The instrument positions are midway between staff positions on the ground, and are similarly entered on the form, which makes it easy to follow.

In the example shown, measurement starts from a Bench Mark. The staff is positioned at station *A*, alongside the Bench Mark: the graduation opposite the Bench Mark is entered in the form in brackets, as it is not an instrumental reading. The altitude of *A* is calculated, as the Bench Mark figure minus the staff reading. The level is set up between *A* and *B*, and a back sight is taken to *A*. The instrument altitude is the altitude of *A* plus the back sight reading to *A*, and it is entered between the lines for *A* and *B*. The foresight to *B* is entered on the line referring to the staff at *B*. The altitude of *B* must be the instrument altitude minus the foresight reading. Stadia readings are also entered with each foresight and back sight. Mental arithmetic will test that the intervals (Upper Stadia–Crosswire) and (Crosswire–Lower Stadia) are equal, or are within 0.005 m. The tacheometric distances 100 x (Upper Stadia–Lower Stadia) are also entered as a check on comparative lengths of foresight and back sight from each instrument position.

Heights can be fixed by one sight only, such as the two intermediate points, Peg 1 and Peg 2, whose altitudes are required for later work. Pegs

STAFF STATION	Back Sight	Inter Sight	Fore Sight	Diff in Altitude	Uncorrected Altitude	Corrtn	Corrected Altitude	STAFF STATION	Remarks	Telescope Altitude	Distance	Upper Stadia	Lower Stadia
BM	(0.783)								Bench mark 167.308 is opposite staff mark of 0.783. Staff base is at 167.308 − 0.783 = 166.525				
A	1.310						166.525	A			160.0	2.110	0.510
									Instrument	167.835			
B	0.870		1.485		166.350	−0.005	166.345	B			159.5	2.280	0.655
											44.5	1.100	0.653
									Instrument	167.220			
Peg 1		2.935			(164.285)	(−0.005)	(164.280)		Traverse peg			3.150	2.720
Peg 2		3.520			(163.700)	(−0.005)	(163.695)		—do—			4.120	2.925
											45.0	2.340	1.890
C	1.320		2.115		165.105	−0.005	165.100	C			120.0	1.920	0.720
									Instrument	166.625			
D	2.550		2.095		164.330	−0.010	164.320	D			120.0	2.695	1.695
											56.0	2.830	2.270
									Instrument	166.880			
Peg 3		1.880							165 metre contour Tape at 1.880		157.0	2.665	1.095
Peg 4		1.880							—do—		128.5	2.525	1.240
Peg 5		1.880							—do—		63.0	2.195	1.565
Peg 6		1.880							—do— (taped at 17.2 metres)		79.0	2.275	1.485
Peg 7		1.880							—do—		56.0	1.130	0.570
E	2.985		0.850		166.030	−0.010	166.020	E			44.0	3.205	2.765
									Instrument	169.015			
F	1.675		0.260		168.755	−0.015	168.740	F			44.5	0.480	0.035
											73.0	2.040	1.310
									Instrument	170.430			
Peg 8		0.430							170 metre contour Tape at 0.430		86.0	0.850	0.010
Peg 9		0.430							—do—		51.0	0.695	0.175
A			3.890		166.540	−0.015	166.525	A			73.0	4.285	3.525
	Σ = 10.710	Σ = 10.695											

Fig. 70   The booking and calculation of a line of levels. Note:

(1) Each *staff* position has a line to itself. If it is at a station in the line of levels, it will have taken to it a Fore Sight and a Back Sight. If an intermediate position, it will receive one reading only.

(2) Each *instrument* position has a line to itself, with the telescope altitude entered, midway between the lines referring to the staff positions on either side, e.g. from the instrument position with altitude 167.220 the surveyor observed a Back Sight of 0.870 to *B* (line above), two intermediate sights to Pegs 1 and 2, and a Fore Sight of 2.115 to *C* (below).

(3) Stadia information is entered on the right. A mental calculation checks that U. Stadia − Crosswire = Crosswire − L. Stadia to within 0.005 m. The distance from instrument to staff, calculated as 100 . (U. Stadia − L. Stadia), is entered as a check on approximate equality of length of sights from the level.

(4) No stadia readings need to be taken for contour points, but if the level is being used in conjunction with a plane table for immediate plotting of contour points, the distances may be needed as shown (see p. 95). The height above ground at which the tape is placed on the staff for contour sighting is readily calculated from the telescope altitude.

(5) At the end of the work, the line of levels closes on a known point. In this example there is a misclosure of 0.015 m, which is distributed through the line on a basis of equal error at all positions.

(6) As a check on arithmetic, Σ Back Sights − Σ Fore Sights = difference in uncorrected altitude between starting and finishing positions of the staff. In this case the difference is the misclosure, as the line closes back to the starting point.

3–7 are points on the 165-metre contour. At the instrument position between *D* and *E*, the telescope height is 166.880 m. Any staff position which gives a reading of 1.880 will be sited on the 165-metre contour. A piece of Scotch tape is attached across this value, and Pegs 3 to 7 represent points at which the crosswire lies across the tape (see p. 92 for a description of contouring). The levelling continues back to A, fixing some 160-metre points en route, with the tape at appropriate graduations. Note that contour points are always Intermediate Sights. The work shows a closing altitude for A slightly different from the starting altitude. This is checked by comparing the sum of the Foresights with the sum of the Back Sights. If the misclosure is confirmed, it can, if small, be adjusted through the

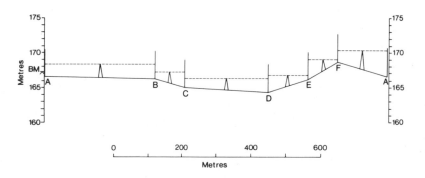

Fig. 71   The profile corresponding to the observations set out in fig. 70. Contour pegs are not shown.

traverse by Bowditch's Rule (see p. 63). The corrections are too small to affect the contouring, but the values for Pegs 1 and 2 are amended, as they will be used later.

The form is easy to follow. It also provides checks on the observations and on most of the arithmetic. It does not check the tape heights as fastened on the levelling staff, and these must always be set out with care. The assistant with the staff counts his paces from the back station to the instrument and walks on a similar number to the fore station, in order to preserve approximate equality of length of sight from each instrument position.

Another method of booking is that known as the "rise and fall" method. The observed readings are as before, but the form of booking and calculation is different. Over each section of the traverse, from staff position to staff position, the back sight reading is subtracted from the foresight reading (fig. 72) to give the height of the fore staff position relative to the back staff position. (Stadia readings can be included as before, as a check: a professional surveyor would usually omit them, but a non-specialist is advised to take them.) Calculation of staff position heights is perhaps easier than in the first method, and is easily checked overall. The sum of the back sight readings minus the sum of the foresight readings should equal the difference in height between the first and last staff positions. For carrying heights through a lengthy traverse, as from a Bench Mark on firm ground to a point across several miles of marsh, this is probably the most convenient form of calculation, but for contouring the first method is less susceptible of error for the non-specialist.

SURVEY: *Back Heath*   OBSERVER: *J. Lloyd*   LEVELLING

DATE :   *8. 10. 73*

STAFF STATION	Back Sight	Inter. Sight	Fore Sight	Diff. in Altitude	Uncorrected Altitude	Corrtn.	Corrected Altitude	STAFF STATION	Remarks	Telescope Altitude	Distance	Upper Stadia	Lower Stadia
BM	(0.783)								Bench mark 167.308 is opposite staff reading of 0.783 Staff base is at 167.308 − 0.783 = 166.525				
A	1.310						166.525	A			160.0	2.110	0.510
B	0.870		1.485	−0.175	166.350	−0.005	166.345	B			159.5 / 44.5	2.290 / 1.100	0.485 / 0.655
Peg 1		2.935							Traverse peg			3.150	2.720
Peg 2		3.520							—do—			4.120	2.925
C	1.320		2.115	−1.245	165.105	−0.005	165.100	C			45.0 / 120.0	2.360 / 1.920	1.890 / 0.720
D	2.550		2.095	−0.775	164.330	−0.010	164.320	D			120.0 / 56.0	2.695 / 2.830	1.495 / 2.270
E	2.985		0.850	+1.700	166.030	−0.010	166.020	E			56.0	1.130	0.570
F	1.675		0.260	+2.725	168.755	−0.015	168.740	F			44.0 / 44.5	3.205 / 0.880	2.765 / 0.635
A			3.890	−2.215	166.540	−0.015	166.525	A			73.0 / 73.0	2.040 / 4.255	1.810 / 3.525
	Σ 10.710		Σ 10.695										

Fig. 72   The "rise and fall" method of booking for the same observations as in fig. 70, but without the contour pegs. Standard procedure is to have two columns entitled Rise and Fall respectively, in place of the single column here called Difference in Altitude. The difference in altitude is calculated from staff position to staff position, so that the back reading to A (1.310) is considered with the forward reading to B (1.485) and it is seen that B is 1.485 − 1.310 = 0.175 m *lower* than A. Similarly, the Back Sight to B (0.870) is considered with the Fore Sight to C (2.115) and C is seen to be 2.115 − 0.870 = 1.245 m lower than B.

As before, the sum of the Fore Sights minus the sum of the Back Sights will give the altitude of the final staff position relative to the starting point, and any small misclosure can be adjusted, provided that it does not exceed 0.020 m per kilometre (or 0.030 m in rough country).

Stadia readings are still taken as a check on Fore Sight and Back Sight readings.

The method is a standard one for professional surveyors, but if used by non-specialist scientists is more likely to lead to errors in contouring than is the method shown in fig. 70.

The theodolite

The theodolite is an instrument designed to measure horizontal and vertical angles. Certain features are essential for this purpose, and must be incorporated in the design.

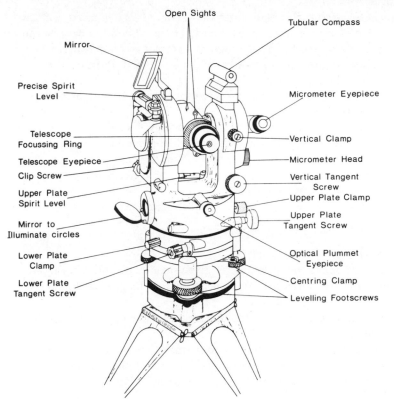

Open Sights

Mirror

Tubular Compass

Precise Spirit
Level

Micrometer Eyepiece

Telescope
Focussing Ring

Vertical Clamp

Telescope Eyepiece

Micrometer Head

Clip Screw

Vertical Tangent
Screw

Upper Plate
Spirit Level

Upper Plate Clamp

Mirror to
Illuminate circles

Upper Plate
Tangent Screw

Lower Plate
Clamp

Optical Plummet
Eyepiece

Lower Plate
Tangent Screw

Centring Clamp

Levelling Footscrews

Fig. 73   A sketch of a modern theodolite. Compare the controls with the highly diagrammatic working sections of fig. 74.

Simple horizontal angular measurement requires:

(a)   a *horizontal graduated circle*

(b)   a *levelling device* to bring the graduated circle into the horizontal
plane

(c)   a *vertical axis* passing through the centre of the circle, and about
which the telescope may be turned from the direction of one object
to the direction of another. This axis need not exist as a component
part: usually the telescope is mounted in such a way that the axis
exists in the design but not as a solid part

(d)   a *centring device* to ensure that the vertical axis is directly over the
station mark. This must include a *clamping screw* to fasten the
instrument in the correct position

(e)   a *horizontal axis* about which the telescope may be moved in a vertical plane, since the points sighted may be at different altitudes

(f)   a *horizontal index mark* connected with the telescope and moving
against the graduations on the horizontal circle, so that a horizontal

rotation of the telescope about the vertical axis will cause a change of reading by the index mark on the circle. The index may be the zero mark of a measuring device such as a vernier or a micrometer

(g)   a *clamping screw* to hold the index at a selected point on the horizontal circle, and

(h)   a *tangent screw* (slow-motion screw) to permit fine adjustment around this selected point

(i)   a mounting of the horizontal circle which will enable it to be turned around the vertical axis so that any selected graduation can be brought to lie in a certain direction from the centre. If the circle is free to move in this way, a *clamping screw* is necessary to fasten it in the desired position, and

(j)   a *tangent screw* will permit fine adjustment of this position.

Measurement of vertical angles requires:

(k)   a *vertical graduated circle*

(l)   a *vertical index mark* connected to the telescope and rotating with it around the horizontal axis against which the graduations on the vertical circle may be read

(m) a *clamping screw* to hold the index at a selected point on the vertical circle, and

(n)   a *tangent screw* to permit fine adjustment around the selected point

(o)   a *sensitive spirit level* to define the horizontal plane, and

(p)   an *adjusting screw* for fine definition of this plane.

Every theodolite will incorporate these features, and an observer faced with an unfamiliar instrument should look for and identify all the components. To assist him, it is normal for manufacturers to pair the clamping and tangent screws (as there are usually three of each); a clamping screw will have the same form as the corresponding tangent screw, i.e. circular, square, triangular, with milled edge or ribbed edge, and the clamping screw is normally smaller than its companion tangent screw. The clamping screw is radial to the horizontal axis (for screws (g) and (i)) or to the vertical axis (for screw (m)), and, as the name implies, the corresponding tangent screw will be at right angles to the radial clamp. The pairing of the screws enables identification by touch as well as by sight: geographers will usually not be working in the dark, but surveyors need this ability to work by finger sensing if observing underground or if taking star observations.

Obviously other features of the theodolite will be expected:

(q)   a *telescope* with open sights on its upper and/or lower surface

(r)   an *eyepiece* for focussing crosswires in the telescope

(s)   a *focussing screw* for obtaining a clear image of the object observed

(t)   *reading devices* for horizontal and vertical circles. These may be verniers or micrometers; if the latter, one or more *micrometer screws* will be present on the instrument. If the micrometer system is internal, one or more illuminating windows will be required to let light into the system, and a *reading eyepiece* will be necessary.

The components of the theodolite may be considered in associated groups or systems, and if the functions of these systems are properly understood, the controls and the operation of the theodolite will present no difficulties.

The tripod

This is entirely separate from the instrument proper, which screws on to it or is clamped on to it. The legs are preferably non-telescopic, firmly constructed from strong wooden parts, with pointed metal shoes at the leg tips. The shoe is fastened to the leg with three screws, and is shaped to provide a "step" which may be used for pressure into the ground by foot action: when pressing on the shoe, the observer's leg should always be inclined parallel to the tripod leg, in order to avoid strain on the shoe and in order to press the leg into a small, fitting hole instead of downwards into a larger open hole. The tripod legs are attached at their upper ends to the tripod head by bolt hinges with butterfly nuts. It is important that the instrument should stand on a firm foundation, and the tripod must therefore be free of "shake", which arises from undesired movement either of the shoes or of the legs relative to the tripod head. Shoes should be checked periodically and the screws tightened if necessary. The bolt hinges between legs and tripod head need a finger check at each station, as the act of placing the tripod legs together while the theodolite is moved from one station to the next will slacken the hinges slightly. The butterfly nuts should not be overtightened, as otherwise the tripod cannot be closed without straining of the wooden legs.

The top of the tripod is protected by a cover which has to be removed before the theodolite can be attached. Most modern theodolite tripods have a small bracket on the underside of one leg, on to which the cover is fitted when first removed. There is usually also a small circular spirit level on the tripod head: this has no particular significance, but assists the observer in setting up the tripod with the head reasonably level. As will be seen, the levelling device on the theodolite compensates dislevelment of the tripod head provided that this is not too great. The circular spirit level keeps the dislevelment within reasonable bounds.

The centring device

This may be part of the theodolite, or part of the tripod. Essentially it consists of two smooth plates, of which one is the base of the theodolite and the other is the top of the tripod. The former is slid over the latter until the plumb bob, suspended by a hook from the bottom of the vertical axis, hangs exactly over the station mark on the ground, when the plates are clamped together. In some patterns, the theodolite is attached to the tripod head by a screw which passes up through the theodolite base plate

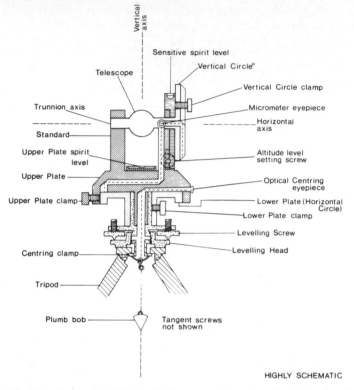

Fig. 74 A very simplified and diagrammatic section through a theodolite. It is intended to show
(a) how each movable screw affects the instrument;
(b) the very approximate alignment of internal optical systems. In this connection note that the vertical circle is shown as being read at its lowest point: in practice it is read against a horizontal index.

and threads into the vertical axis (see fig. 74). The plumb bob is hung from the bottom of this screw. Slight slackening of the screw allows the theodolite and its base plate to be moved laterally in any direction; tightening of the screw clamps the base plate to the tripod head, using beneath the tripod head a slide bar which can swing sideways until locked by the grip of the screw from beneath and the instrument and tripod head from above. With this pattern, the whole of the theodolite moves sideways on the tripod head.

In other patterns, the theodolite threads firmly on to the tripod head, and the lower part of the theodolite, i.e. the levelling head, remains firmly fixed while the upper part of the instrument slides laterally; in this case the centring base plate is part of the levelling head instead of the tripod head, and the plumb bob, hanging directly from the theodolite, passes, through the tripod head. The clamp in this pattern may be a single screw or may be a locking ring with or without projecting lugs to provide a finger hold.

Lateral movement in all cases rarely exceeds 25 mm. It is therefore important that the tripod should itself be approximately centred. On even ground, the legs should be set at intervals of about 120°, and equidistant from the centre mark. On uneven ground, it is best to have two legs at approximately the same altitude and equidistant from the mark, with the third leg on the steepest slope above the station mark. If the third leg is sited first, the other two may be held clear of the ground until the plumb bob is over the mark and the tripod head is approximately horizontal: then the two free legs are lowered until they touch the ground. All three legs are then pressed home to give a firm base before accurate centring is started. If setting up the tripod before the attachment of the theodolite, remember that the latter adds about a third of a metre to the overall height. The telescope should be at a comfortable eye level in the finished position, neither too high nor too low.

The plumb bob used in centring is suspended on a cord with a button slide so that its height above the ground can be varied. In windy conditions the plumb bob may swing wildly unless screened during centring, and modern theodolites usually have an optical centring device for such conditions, consisting of a small horizontal eyepiece and a 45° prism on the vertical axis to direct the line of sight downwards; the image of the ground mark can be focussed by turning or sliding the small eyepiece. With this device, the centring clamp is released and the instrument slid laterally while the observer brings the image of the ground mark to the centre graduation of the eyepiece when the clamp is tightened. It is most important that if an optical centring device is used, levelling of the instrument must be completed *before* centring is attempted, as otherwise the line of sight is not vertical below the axis. Centring should always be accurate to within 2 mm, as otherwise short lines will introduce unacceptably large angular errors into the work (see p. 78).

The levelling head

This consists essentially of two parallel rings (see fig. 74). The lower ring is threaded on its inner side to screw on to the tripod head. The upper ring holds the bushes in which are fitted the three levelling screws, these screws having their lower ends resting on the lower ring. A socket joint permits movement of the upper ring relative to the lower.

The levelling head carries above it the remainder of the theodolite, fitting together and moving as a whole. On the instrument is a spirit level used in putting the horizontal circle into the horizontal plane. To bring the bubble to the centre of this spirit level, the footscrews of the levelling head are rotated. The level tube is set parallel to two footscrews, and these are turned in opposite directions by equal amounts In fig. 74, if the left-hand screw is rotated in an anti-clockwise direction it will be screwing further into the upper ring and will bring the left side of this down towards the tripod head, while clockwise rotation of the right-hand screw will raise the ring on that

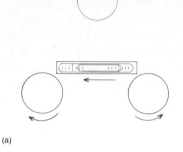

(a)

(b)

Fig. 75   (a) The Upper Plate is turned until the spirit level is parallel to two foot screws. These are turned in *opposite* directions by equal amounts, when the bubble will move in the same direction as the left thumb of the surveyor.
(b) After the bubble has been centred in stage (a), the Upper Plate is turned through 90°, and the third footscrew is used *alone,* to bring the bubble to centre.
Both (a) and (b) operations are then repeated.

side. This tilts the whole of the instrument above it. When the bubble is central, the instrument is turned through a right angle, and the third foot-screw is employed to bring the bubble central in the new position. The instrument is returned to its first position and any slight dislevelment of the bubble is rectified as before by equal and opposite rotation of the two footscrews: it is then again turned at right angles and the third screw is used to make a final adjustment. The bubble should now remain in the centre of the tube irrespective of the direction in which the telescope is pointing. If it does not, the spirit level will need to be adjusted as described on p. 142.

The bubble will move in its tube in the same direction as the observer's left thumb when turning the screws (see fig. 75).

The lower plate system

This is a major part of the theodolite. It consists of the horizontal plate carrying the horizontal circle, and mounted on the outer axis. A vernier theodolite has the horizontal circle graduated on silver and mounted at the bevelled edge of the plate. A theodolite with an internal optical system has a graduated glass circle totally enclosed within a metal case: the circle is illuminated by means of a small mirror which can be set to reflect light through a ground glass window into the instrument, whence the circle image is reflected by a system of lenses and mirrors to a special eyepiece and micrometer.

The lower plate system can be fastened by the clamping screw. If this is tight, it fastens the lower plate and outer axis to the tangent screw. The tangent screw is threaded through a lug projecting from the levelling head, so that when the clamping screw is tight, the lower plate is immovably fixed

(a)

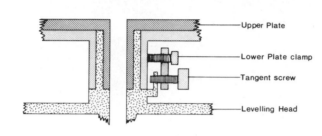

(b)

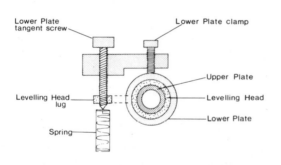

HIGHLY SCHEMATIC

Fig. 76   Highly diagrammatic sketches of tangent screw operation. (a) and (b) represent the lower plate controls: the clamp screw fastens the lower plate to the block holding both clamp and tangent screw, and is itself held to the levelling head by the tangent screw. If the tangent screw is turned, the block through which it is threaded will rotate laterally either towards the point of the screw or towards the head, moving with it the lower plate which is clamped to it.

to the levelling head. However, rotation of the tangent screw causes a slow rotation relative to the lug, and so relative to the levelling head, of the tangent screw, the clamp screw, and the lower plate system. Figure 76 shows the tangent screw in a highly diagrammatic form. It should be clear that if the clamping screw is not tight, rotation of the tangent screw will move only the casting holding the two screws, without gripping the outer axis and the lower plate. This must always be true of a tangent screw: there

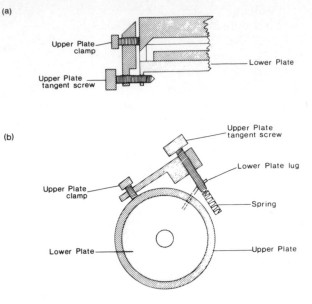

HIGHLY SCHEMATIC

Fig. 77   (a) and (b) show in highly diagrammatic form the tangent screw system of the Upper Plate.

Note that as with fig. 76 if the tangent screw is turned too far it will either come to a stop by bringing the block against the lug or will lose contact with the lug against which it should bear. Note also that the tangent screw cannot turn the plate unless the clamp is tight.

A rather similar tangent screw operates between the vertical circle and the sensitive level frame.

must be prior tightening of the clamping screw if the tangent screw is to operate.

Some modern theodolites do not have a clamp and tangent screw for the lower plate. Instead there is a single milled wheel which will rotate the lower plate in continuous gearing. This milled head is concealed by a metal cover, and as long as this is kept closed no confusion is possible between upper plate and lower plate controls. Although this is convenient for many purposes, the design does not easily allow some operations, such as repetitive observation of an angle in subtense measurement (see p. 22).

The upper plate system

The upper plate rests on the lower plate and is attached to the inner axis. The clamping screw threads through the upper plate and when tight binds it to the tangent screw. This is threaded through a lug from the lower plate, and when rotated will move the whole of the upper plate system relative to the lower plate. An index mark rests against the graduated circle. If it is a vernier theodolite, there will be two verniers on opposite sides of the circle, each with an index mark; both verniers will be read and the mean value taken for the minutes and seconds, the degree figure always being taken from the same vernier. If it is a micrometer theodolite, two micrometers replace the verniers. If it is a theodolite with an internal optical system, a single micrometer reads one side, or the two opposite sides, of the circle, depending on the design.

The upper plate carries on it the spirit level and the standards which support the telescope system.

The vertical reading frame

This is essentially a T-shaped bracket adjacent to one of the standards. The horizontal arm carries one or two index marks. As with the horizontal circle, these indices may be verniers, micrometers, or part of an internal optical system. The line joining the two index marks (or one index and the centre of the trunnion axis) defines the horizontal plane for purposes of vertical angular measurements, and it is obviously important that this should be correctly adjusted. The sensitive spirit level is used in the definition of the horizontal plane, and the axis of this spirit level should be parallel to the line joining the index marks. The reading arm and spirit level are connected to the upper plate system by means of a vertical arm through the end of which is threaded a clip screw which bears upon one of the upper plate standards. Rotation of the clip screw will clearly move the whole of the vertical reading frame and the telescope and vertical circle system if that is fastened to the vertical frame. It is clear that once the line joining the indices has been set parallel to the axis of the sensitive spirit level, both can be brought to the horizontal position by use of the clip screw. Once this has been done, alteration of the clip screw will result in dislevelment.

Some theodolites have a sensitive spirit level with graduations on the glass tube, so that dislevelment at the time of observation can be adjusted in

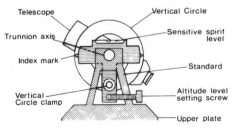

HIGHLY SCHEMATIC

Fig. 78   A very simplified and diagrammatic section of the vertical system of a theodolite. Note that
(a) the index is stationary upon the frame, and the vertical circle rotates with the telescope;
(b) movement of the altitude level setting screw (sometimes called the clip screw) will rotate the index frame and the sensitive spirit level about the trunnion axis. The design of this screw in fact is very different from the simple sketch given here.

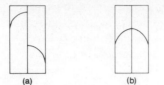

Fig. 79   The mirror image of a split bubble:
(a) showing dislevelment;
(b) with the two ends brought into balance.

terms of the graduation readings of the two ends of the bubble. Others have
angled prisms and a mirror which reflects the images of the two ends of the
bubble. If there is dislevelment, the two ends are not in correspondence, as
seen in fig. 79a; slight adjustment with the clip screw results in a united
image as in fig. 79b.

The telescope and vertical circle

The telescope is mounted on the trunnion axis which is held between the
two standards of the upper plate. The telescope has attached to it the
vertical circle which rotates with it. In this respect the vertical circle differs
from the horizontal circle. The latter is fixed, but the reading indices move
round the circle as the telescope is turned laterally. The vertical reading
indices are fixed in the vertical frame, and the vertical circle moves past
them as the telescope is rotated about the trunnion axis. A clamping screw
fastens the telescope and circle to the vertical tangent screw (not shown)
which moves telescope and circle.

The telescope has an adjustable eyepiece, for the focussing of crosswires,
and an adjusting screw for focussing the object sighted. Sometimes this is a
large screw on the outside of the telescope, sometimes it is a milled ring
which forms part of the telescope casing.

Observations with the theodolite are stated to be Face Left when the
vertical circle is on the left of the telescope as the observer views the target
object, and Face Right when it is on the right. Instruments used by geographers will normally be transit theodolites, a transit being an instrument
in which the telescope may be turned through 180° about the trunnion
axis without removal from the standards, i.e. the height of the trunnion axis
above the upper plate exceeds the length of the telescope from axis to either
end.

The graduations of the vertical circle may follow one of several patterns.
Some instruments read 0° at the horizontal, on either face, and 90° at the
vertical, up or down. A common method is graduation from 0° at the horizontal on Face Left, through 90° at the zenith to 180° on the horizontal,
Face Right, and on through 270° to 359° and 0°. A reading of 0° on Face
Left or of 180° on Face Right indicates the horizontal, and any value between
0° and 180° represents an angle of elevation; between 180° and 360° is an

angle of depression. This is, of course, different from the pattern of readings obtained on the two faces with the horizontal circle. A *FL* reading of the horizontal circle to station *X* gives, for example, 136° 52′ 40″, and the *FR* reading to the same station should differ by 180°, i.e. it should read 316° 52′ 40″. When the vertical circle is graduated from 0° to 360°, vertical readings on *FL* and *FR* will not differ by 180°, as is clear from fig. 80. A

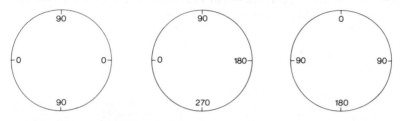

Fig. 80   Vertical circles can be graduated in a variety of ways, of which three are illustrated. Before starting observations with a theodolite it is essential to check on the characteristics of the instrument.

FL reading of 03° 52′ 40″ (representing an angle of elevation) will be matched by a *FR* reading of 176° 07° 20″; a depression *FL* reading of 356° 11′ 50″ would be matched by a *FR* reading of 183° 48′ 10″. Instruments graduated upwards from 0° at base through 90° at the horizontal to the zenith at 180°m on both faces, should give identical vertical readings on the two faces. The pattern of vertical circle graduation should be checked on first taking out an unfamiliar theodolite.

Reading devices

A common pattern of theodolite has a vernier reading to 20″ of arc. A vernier theodolite usually has four verniers, *A* and *B* on opposite sides of the horizontal circle, and *C* and *D* at each end of the horizontal arm on the vertical bracket, reading the vertical circle. Degrees are recorded as for *A* and *C* respectively; minutes and seconds are meaned from *A* and *B* and from *C* and *D* respectively, e.g.

$$A = 172° 31′ 00″ \qquad C = 03° 22′ 00″$$
$$B = 352° 30′ 20″ \qquad D = 183° 21′ 20″$$
$$\text{Mean} = 172° 30′ 40″ \qquad \text{Mean} = 03° 21′ 40″$$

It will be appreciated that as *A* and *B*, or *C* and *D*, are at opposite ends of a diameter, their readings should show a difference of approximately 180°. It will also be realised that a Face Left mean vertical angle may be 03° 21′ 40″ and the corresponding Face Right mean may be 176° 38′ 40″. The mean of the two will be

$$\frac{03° 21′ 40″ + 03° 21′ 20″}{2} = 03° 21′ 30″.$$

A more refined method of measurement is the micrometer. This consists of a screw which moves the image of the circle under a crosswire or crosswires (or vice versa) in such a way that one rotation of the screw head exactly moves the image from one circle graduation to the next. Thus a circle graduated at 10′ intervals might have a micrometer head subdivided into 120 sections, each representing $\frac{10'}{120} = 5''$ of arc. Micrometers are easy to read and more refined than verniers in their graduation, unless the vernier is made inconveniently long. Micrometer patterns vary, but an individual

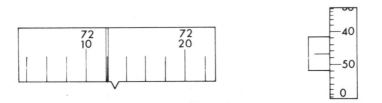

Fig. 81   The left-hand illustration shows the image in the micrometer eyepiece. In this case the notch marks the exact reading, and rotation of the micrometer head (shown on the right) moves the double line across the image to stand astride a graduation of the main scale. Two rotations of the micrometer head move the double line from one graduation to the next.
The two numbered graduations are for 72° 10′ and 72° 20′, so that the notch is between 72° 12′ and 72° 14′. The micrometer head reads 47″ and from the position of the notch it is clear that the reading is 72° 12′ 47″ and not 72° 13′ 47″.

micrometer can easily be understood. For example, a micrometer might be graduated from 0 to 60, with 2′ graduations on the main scale. If it takes two revolutions to move the index image from one graduation to the next, it is clear that one revolution represents one minute and the micrometer readings are in seconds; the user should know which minute reading to enter in the field book. In fig. 81 there should be no doubt as to whether the correct reading is 72° 12′ 47″ or 72° 13′ 47″. The micrometer wheel has been turned until the thin double line stands astride one graduation of the main scale; the micrometer index then reads 47″; the index notch on the main scale indicates the exact position on the scale, and cannot be anything other than 72° 12′ 47″. In this case revolution of the micrometer moves the double line across the scale image. If when the crosswires are aligned on a scale graduation the notch position does not agree with the micrometer reading, the micrometer is out of adjustment and requires professional attention.

Theodolites with internal optical systems use a variety of micrometer devices, and there is no standard routine for the determination of the correct reading. In some micrometers, turning of the micrometer head (which is not itself graduated) moves the circle image past a fixed mark, and the observer turns the micrometer head until one graduation intersects the mark, when an illuminated window gives the micrometer reading. Figure 82 illustrates this pattern, (a) immediately after intersection of the target object, and (b) after rotation of the micrometer head. Note that only one graduation can, as a general rule, be intersected, as the micrometer run is limited in this case to a little over 20′, with the main scale having 20′ graduations. In fig. 82, (c) and (d) illustrate the special case when either of two graduations can be used, as both fall just within the micrometer range. It can be seen that the

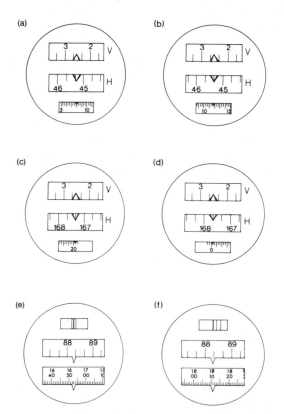

Fig. 82   Micrometer readings in a modern theodolite, graduated to 20″.
(a) As seen after intersection of the target. Neither horizontal nor vertical scale has a graduation bisecting the index.
(b) After rotation of the micrometer head, the horizontal scale image has moved laterally until one graduation is straddled exactly by the index, in this case the graduation for 45′ 20′. The bottom window index is above 11′ 40″ but is less than 12′ 00″: call it 11′ 50″. The two readings added together give 45° 31′ 50″.
(c) The horizontal scale reads 167° 20′ + 20′ 20″ = 167° 40′ 20″.
(d) The horizontal scale reads 167° 40′ + 00′ 00″ = 167° 40′ 20″.
   Micrometer readings in another theodolite, graduated to 1″. With this instrument the eyepiece shows either the *H* or the *V* scale depending on the position of a switch. The surveyor intersects the target with the telescope and then turns the micrometer head until the upper window shows the image of a movable double line standing astride the fixed centre mark.
(e) The centre window shows that the reading will be between 88° 00′ and 88° 20′. The lower window shows 16′ 53″ to give a circle reading of 88° 16′ 53″.
(f) The two movable lines are erroneously both to one side of the centre mark, to give a false circle reading of 88° 18′ 11″.

same reading is obtained irrespective of the graduation used. Some internal micrometers use opposite sides of the circle, giving a double image. Others use a separate "window" which requires rotation of the micrometer head until a fixed line is straddled by a movable pair of lines, or some similar device. In this case care must be taken to ensure that the two moving lines straddle the fixed mark as in fig. 82(e), and are not both spaced to one side of the mark as in fig. 82(f).

Lack of uniformity extends to the controls. In some instruments a single micrometer head operates for both horizontal and vertical circle readings; in other instruments there are separate micrometer heads, and a switch marked *H* or *V* sets the internal system for use of the appropriate control. Some theodolites have a micrometer eyepiece alongside the telescope eyepiece, so that the image of the circle and the micrometer reading can be viewed by a very small sideways movement of the head after the intersection of the target object; others have on one of the standards a micrometer eyepiece which can be turned in either direction to lie in a horizontal position parallel to the telescope, so that it can be read with equal ease on either Face Left or Face Right.

Adjustments of the theodolite

Station adjustments   These are the adjustments necessary at any station preparatory to the taking of readings. They include:

  (a) Centring
  (b) Levelling by means of the footscrews (as on p. 134)
  (c) Focussing of the crosswires by rotation of the eyepiece
  (d) Focussing of the telescope on the target object
  (e) Levelling by means of the sensitive bubble before taking a sighting for a vertical angle (see p. 9).

Semi-permanent adjustments

  (f) Plate levels
       After levelling with the footscrews as described on p. 134, the upper plate should be turned through 360°. If the bubble of the spirit level stays central throughout, the instrument is in adjustment. If it is central when parallel to two footscrews, but off-centre when turned through 180°, bring it half-way back to centre by use of the footscrews and complete the centring by turning the locking nut and adjusting nut (or capstan-headed screws) at one end of the spirit level. (If the movement required is large, loosen the nuts at one end of the spirit level and tighten those at the other. Repeat the process with use of all levelling screws, making further adjustments as necessary, until the bubble remains central in all positions of the upper plate.
  (g) Sensitive level and collimation in altitude
       These adjustments can sometimes be effected independently, and

sometimes need to be combined, depending on the design of the theodolite. The essential requirement is that when the line of sight is horizontal, the sensitive bubble shall be central and the vertical circle shall read zero. Clearly this desired state can be disturbed if the sensitive spirit level is not parallel to the upper plate when this is horizontal. The first stage of the adjustment is therefore to test the sensitive level as for the plate level in (f) above, but taking up half the correction on the clip screw instead of with the levelling screws, the other half being made with the nuts holding the sensitive level. When this has been done the bubble should be central for all directions of the telescope.

The telescope is then set to read zero on the vertical circle. With the bubble central, a reading is taken to a graduated staff at about 100 m distance, on Face Left, and a similar reading is taken with the instrument at Face Right. If there is no error, the two readings are identical. If they differ, as in fig. 83, there is a collimation error, the line of sight not being parallel to the axis of the sensitive level. In this case the clip screw is used to bring the line of sight to the mean of the two readings on the staff; the bubble is then brought back to centre using the nuts at the ends of the spirit level. The centrality of the bubble in all positions should then be retested, using only the levelling screws (not the clip screw) to achieve this. If the collimation error cannot be eradicated, the instrument should be serviced professionally.

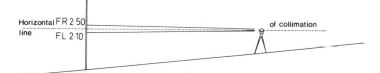

Fig. 83   Adjustment of the theodolite for collimation in altitude.
A graduated staff is held 100 m from the theodolite. With the sensitive bubble of the theodolite at centre, a Face Left reading is taken to the staff, to give 2.10 m; a similar Face Right reading is 2.50 m.
The clip screw is used to bring the crosswire to read 2.30 m on the staff. This causes dislevelment of the sensitive bubble, which is brought back to centre by adjustment of the screw nuts holding the spirit level.

Older types of theodolite allowed independent adjustment for collimation. The horizontal and vertical crosswires in the instrument should intersect at a point on the horizontal axis of the telescope. If either axis is displaced, the line of sight through the intersection will not give a true reading. Vertical and lateral adjustment for the crosswire diaphragm was possible with locking screws between which actual crosswires were stretched. Vertical adjustment for collimation in altitude was achieved with the locking screws, and a similar lateral adjustment could be made for collimation in azimuth.

The latter is tested as follows. A distant mark approximately on the horizontal plane through the instrument is intersected on Face Left and the horizontal circle reading noted. The telescope is then turned to Face Right, the mark intersected and the circle reading noted. The difference between the two readings should be 180° 00′ 00″. If the difference consistently exceeds 1′ with a 20″ theodolite (or 15″ with a 1″ theodolite) it is necessary to have the instrument serviced. The non-specialist is advised not to attempt adjustment for collimation in azimuth with a modern instrument, but should seek professional assistance. The same advice should be followed if, after adjustment of the spirit levels, it is found that the vertical plane of collimation is not vertical. This can be tested by suspending a plumb bob on a long cord from a tree, window, etc., sufficiently close to the theodolite to give elevations of 0°–45° when viewed through the telescope. With the horizontal circles clamped, and the cord intersected by the crosswires at zero elevation, the cord should remain intersected as the elevation is increased, provided of course that the test is made in still air in which the

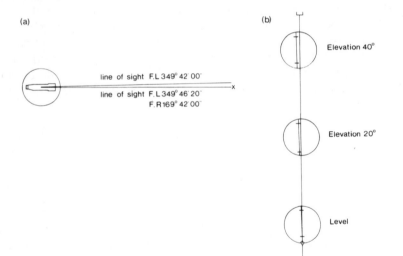

Fig. 84   Two conditions which, if they persist after completion of the adjustments listed in the text, should make the non-specialist seek professional assistance with the servicing of the theodolite.
(a) Intersection of the mark X gives consistent differences between Face Left and Face Right which exceed the allowable limit of 1′ of arc.
(b) With increasing elevation the vertical crosswire diverges from alignment with the vertical cord, notwithstanding centrality of the spirit level bubbles.

cord is motionless. Figure 84(b) illustrates the situation when the plane of collimation needs attention.

(h) General
    There are other adjustments which will be checked by the professional surveyor or engineer, but which in general will not affect the work of the field scientist. The latter should check his instrument for

collimation errors before going to the field, and if these are present
he should have the machine serviced before departure. Once in the
field he should not need to perform the semi-permanent adjustments
unless he is observing for a month or more, when he should check
thoroughly the continued accuracy of his instrument. He should,
however, make a quick check of his instrument each day on first
setting up in the field.

The vertical adjustments given above are particularly important
when a split bubble image is used (see p. 138).

The sextant

The sextant is used much less by landsmen than might be expected. It is
capable of measuring horizontal and vertical angles to 12″ of arc, is no more
difficult to use than is a theodolite, and is easier to transport. It can be used
under conditions which make theodolite observations impossible, e.g.
angular observations from a boat offshore.

It consists of a metal frame carrying part of the arc of a circle (usually
about one-sixth of a circle, hence its name). Against the main scale is a vernier
(see fig. 85) mounted on an arm pivoted at the centre of curvature of the
main scale. A clamp screw and a tangent screw control movement of the
vernier arm. Telescope attachments screw into a bracket on one arm of the
main frame, aligned with a glass plate, half silvered, on the other arm of the
frame. A series of dark glasses may be turned up behind the horizon glass or
in front of the index glass.

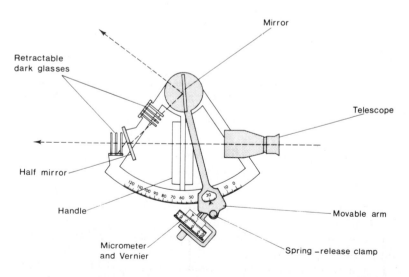

DIAGRAMMATIC

Fig. 85   The component parts of the sextant.

For astronomical vertical angles at sea, the observer looks through the plain section of the horizon glass at the sea horizon, and brings the image of a star, or of the sun's upper or lower limb, into the mirrored part of the horizon glass, aligned with the horizon. This is done by moving the index arm, clamping when the image comes into view, and completing the sighting with the tangent screw. For observations without an available free water surface, an artificial horizon is used, preferably in the form of a dish with a free mercury surface. In this case, the observer views the reflection of the sun through the clear glass and brings the image in the mirror down to touch the image from the artificial horizon: as the latter has been reflected only once while the former has been reflected twice, the images appear to be travelling in opposite directions.

Some modern sextants have only one control knob, which is the tangent screw shifting the movable arm over the scale and is also a micrometer head graduated in minutes, with a vernier to read to 12 seconds of arc. The clamp holding the arm against the scale has a spring action which is released by a firm grip so that the arm can be swung freely. When the desired image is reflected in the field of view the spring is released and the arm automatically clamps.

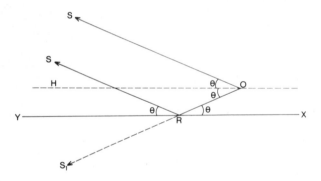

Fig. 86   The observer uses the sextant to view the sun (or star) directly and to measure the altitude angle above the horizon. At sea the true horizon may be clearly defined, but on land it is not so, and an artificial horizon R is used. The angle $R\hat{O}S$ between the reflected and observed images is then twice the altitude angle required.

Figure 86 illustrates the use of the artificial horizon. The observer O wishes to observe the angle $H\hat{O}S$, OH being the horizontal, but H cannot be defined. The artificial horizon at R reflects the sun or star so that $Y\hat{R}S = O\hat{R}X$, XY being the horizontal defined by the mercury surface. $R\hat{O}S = 2Y\hat{R}S$, so that the observed angle $R\hat{O}S$ must be halved before calculations are attempted. As with the theodolite, sun observations should be paired to use the upper and lower limbs.

Figure 87 indicates the geometry of observation to the sea horizon. OH is the horizontal line of sight through the clear glass, and HI is the reflected sight to the index mirror I. I is turned until the sun or star S comes into

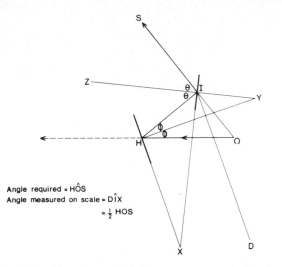

Angle required = HÔS
Angle measured on scale = DÎX
                        = ½ HOS

Fig. 87   The geometry of the measurement of altitude using the sea horizon. See the text for the proof.

view against the horizon at H. The angle required is $H\hat{O}S$. The angle measured by the instrument is $D\hat{I}X$ (i.e. the angle between the planes of the two mirrors: when the I mirror is parallel to the H mirror, X coincides with D and the index arm reads zero on the graduated scale).

Let HY and IY (IZ) be the perpendiculars to the two mirrors. As the angle of incidence equals the angle of reflection, we can call the angle $H\hat{I}S = 2\theta$ and the angle $I\hat{H}O = 2\phi$.

The angle $H\hat{X}I = D\hat{I}X =$ the angle of movement of the index arm over the graduated scale. However, $H\hat{X}I =$ the angle between the two mirrors = the angle between the perpendiculars to the two mirrors = $H\hat{Y}I$.

In $\triangle HIY$

$$I\hat{H}Y + H\hat{Y}I \ = H\hat{I}Z = \theta$$
$$H\hat{Y}I \ = \theta - I\hat{H}Y = \theta - \phi$$
$$\therefore \quad D\hat{I}X \ = \theta - \phi$$

In $\triangle HIO$

$$I\hat{H}O + H\hat{O}I \ = H\hat{I}S = 2\theta$$
$$H\hat{O}I \ = 2\theta - I\hat{H}O$$
$$= 2\theta - 2\phi$$
$$\therefore \quad H\hat{O}I \ = 2D\hat{I}X$$

i.e. the angle $D\hat{I}X$ moved by the index arm is only half the angle $H\hat{O}S$ which is required. For this reason the main scale is graduated so that 45° of scale is marked with 90° in angular divisions, so that the reading taken from the scale is correct for the angle required.

It should be clear that if the sextant is turned on its side, so that the graduated arc is horizontal, the instrument can be used to measure horizontal angles between two points. If point A is viewed through the clear section of the horizon glass, and the image of point B is made to coincide with it, the angle at the observer subtended by the two points A and B can

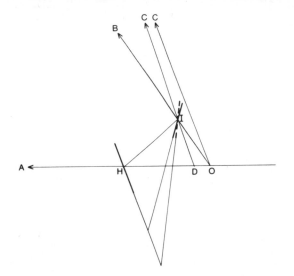

Fig. 88   The observer measures the angle $A\hat{O}B$ with the movable mirror at I in the position of the broken line. If he then wishes to measure the angle $A\hat{O}C$ he turns the mirror at I into the position of the solid line, when the image of C is made coincident with A. It appears that $A\hat{O}C$ does not equal $A\hat{D}C$, which is measured by the sextant. The angle between the two mirrors will be half the required angle (see fig. 86) irrespective of whether the observer is at D, O or any other point on AH produced (see text). This means that the position of the surveyor in relation to the points being observed will not vitiate results.

be measured. Strictly speaking, this is not the angle $A\hat{O}B$. Figure 88 shows that for the angle between the rays to points A and B, the angles $H\hat{I}B$ and $I\hat{H}O$ will determine the value of $A\hat{O}B$. If, however, point C is considered, the angles $H\hat{I}C$ and $I\hat{H}D$ will determine the value of $A\hat{D}C$, which is not quite the same as $A\hat{O}C$. What is measured by the scale is the angle $A\hat{O}C$, i.e. the angle between the mirrors, which will give the angle at the observer whether he is at O or at D or at any other point on the line AH produced. The only moving part of the instrument is the index arm and the mirror at I, and it is therefore the angle at I which is significant, not the angle at some point, on AH produced, which will vary with the stations being observed.

In fig. 89(a), point A is so far away that HA and IA are effectively parallel. If A is the distant horizon, for example, the angle of elevation will be zero. The two mirrors are therefore parallel and the observed horizon HA and the reflected image IA will coincide.

In fig 89(b), point B is on the line HA but is not far distant from the observer. To obtain coincidence of the image IB with the line of sight HB, the

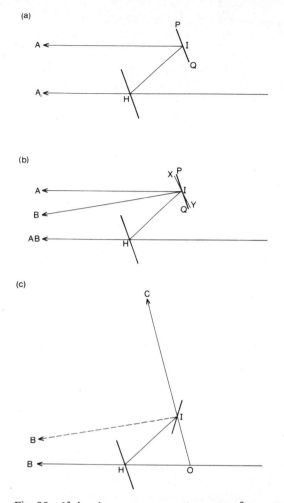

(a)

(b)

(c)

Fig. 89 If the observer measures the angle $B\hat{O}C$, and B is nearer than 2 km, a parallax error greater than $10''$ is introduced because the rays from B to H and I are not parallel.

In (a) when A is distant (e.g. the sea horizon) the two rays from A are parallel, and the mirrors at H and I are parallel, with the sextant scale at zero, when the images are fused.

In (b) it can be seen that the mirror at I is turned from PQ to XY in order to fuse the images of B. The angle recorded on the sextant scale will equal the parallax angle $I\hat{B}H$.

In (c) the measured angle $B\hat{I}C$ differs from the desired angle $B\hat{O}C$ by the parallax angle $I\hat{B}H$.

See the text for a full explanation.

I mirror must be turned from the position PQ to the position XY, and the scale will record an angle. The angle $I\hat{B}H$ is the *sextant parallax*, or the angle subtended at the point B by the index glass and the horizon glass. As the angle of incidence equals the angle of reflection, the I mirror has been turned by $X\hat{I}P$, which $= \frac{1}{2}B\hat{I}A$, in order to turn from the A image at H to the B image at H, and $\frac{1}{2}B\hat{I}A = \frac{1}{2}I\hat{B}H = $ half the sextant parallax. However, as the sextant scale records twice the angular movement of the *I* mirror, on fusing the reflected image of *B* with the observed ray *HB* the scale will give a direct reading of the parallax at *B*.

In fig. 89(c), the observer wishes to obtain the angle $B\hat{I}C$, which has been shown to be the significant angle. However, because the sextant is adjusted to read zero with the *I* mirror parallel to the *H* mirror, as in fig. 89(a) the angle as recorded on the scale is $B\hat{O}C$ and not $B\hat{I}C$.

$$B\hat{I}C = B\hat{O}C + I\hat{B}O$$

$$= \text{measured angle + parallax}$$

An observed value can therefore be corrected by a preliminary observation for parallax, obtained as above, Alternatively, the parallax angle can be calculated:

In ΔIHB

$$\frac{IH}{\sin B} = \frac{IB}{\sin H} = \frac{IB}{\sin(180 - 2\phi)} = \frac{IB}{\sin 2\phi}$$

$$\sin B = \frac{IH \sin 2\phi}{IB}$$

The angle 2ϕ is a constant for the instrument.

If $2\phi = 40°$, for example, and $IH = 80$ mm, it is possible to calculate parallax angles against distance from *B*, such as:

At	10.6	kilometres,	the parallax angle is	01″
	2.12	kilometres,	the parallax angle is	05″
	1.06	kilometres,	the parallax angle is	10″
	530	metres,	the parallax angle is	20″
	354	metres,	the parallax angle is	30″
	177	metres,	the parallax angle is	60″

If angles are required to the nearest 10″, parallax correction will be necessary if the R.O. is closer than 2 km. Whether or not the parallax correction is made depends upon the degree of accuracy with which the survey is being conducted. It follows that in measuring an angle, the R.O. should be as distant as possible. If the angle to be measured is between two points both relatively close to the observer, it is better to observe the angle from a distant point *X* to point *A*, and from *X* to *B*, and to obtain the angle between the rays to *A* and *B* by subtraction.

The sextant usually observes angles up to about 110°. If a larger angle $A\hat{O}B$ is required, it needs to be obtained by the cumulative observation of angles $A\hat{O}P$, $P\hat{O}Q$ and $Q\hat{O}B$, *P* and *Q* being distant intermediate points.

Tilt correction

Of much greater importance than the parallax correction is the correction to be applied to "horizontal" angles observed with the sextant, to adjust for the tilted plane on which the angle is actually measured. A theodolite measures the true horizontal angle at the observer's position, as the gradu-

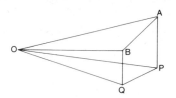

Fig. 90   An observer at *O* wishes to measure the *horizontal* angle $A\hat{O}B$. A theodolite has its horizontal circle in the plane *POQ* and the telescope lies in the planes $O\hat{A}P$ and $O\hat{B}Q$ so that the measured angle is true.

A sextant operates in the plane *OAB*, i.e. it is tilted at the time of observation so that the measured angle is $A\hat{O}B$ and not $P\hat{O}Q$. The correction to the horizontal depends on the size of the observed angle $A\hat{O}B$ and on the angles of elevation or depression $A\hat{O}P$ and $B\hat{O}Q$.

ated circle remains horizontal and angles of elevation or depression from the observer to the target stations are accommodated by vertical movement of the telescope. The sextant measures the angle on the tilted plane defined by the observer's position and the two stations, i.e. on the plane *AOB* in fig. 90. The true horizontal angle is $P\hat{O}Q$, and the relationship between $A\hat{O}B$ and $P\hat{O}Q$ is given by the formula

$$\cos P\hat{O}Q = \frac{\cos A\hat{O}B - \sin h_A \sin h_B}{\cos h_A \cos h_B}$$

where h_A and h_B are the angles of elevation of *A* and *B* above *O*. The correction may be very considerable, e.g. with h_A and h_B both 10° elevation from *O*, a measured angle of 60° needs a positive correction to 60° 01′ 23″. If $h_A = 0°$ and $h_B = 10°$, for an observed angle of exactly 60° there would be a negative correction of 30′ 42″ to give an angle of 59° 29′ 18″. If, under the same conditions of $h_A = 0°$ and $h_B = 10°$, the observed angle were exactly 30°, the corrected horizontal angle would be 28° 25′ 54″. It is important to remember to use a negative value for the sine of an angle of depression.

As horizontal angles are required to an accuracy of 20″ for most survey purposes, this error due to tilting is obviously serious and must be rectified. It can be calculated from the formula given, but if the angles h_A and h_B exceed 10°, or if their difference exceeds 10° (as one or both may be negative angles of depression, below the observer's position), it becomes virtually impossible to observe vertical angles with sufficient refinement to make the correction calculation possible within the limits required. Figure 91 indicates the correction values for observed values of 60°, between limits 0° and +10°. It can be seen that the corrections are small when h_A and h_B are small, and for this reason the sextant is an admirable instrument for hydrographical surveys. Fixed marks at sea level or only slightly above it, with minimal angles of elevation resulting, enable fixations by sextant to be carried out within the plottable error of mapping operations.

h_B \ h_A	0°	-1°	-2°	-3°	-4°	-5°	-6°	-7°	-8°	-9°	-10°
0°	60°00'00"	59°59'42"	59°58'47"	59°57'16"	59°55'09"	59°52'25"	59°49'03"	59°45'05"	59°40'28"	59°35'12"	59°29'18"
+1°		59 58 11	59 56 04	59 53 20	59 50 00	59 46 02	59 41 27	59 36 14	59 30 23	59 23 52	59 16 42
	59 59 42	60 00 36									
+2°			59 52 44	59 48 47	59 44 14	59 39 02	59 33 13	59 26 46	59 19 39	59 11 53	59 03 26
	59 58 47	60 00 54	60 02 25								
+3°				59 43 37	59 37 50	59 31 25	59 24 21	59 16 39	59 08 17	58 59 14	58 49 31
	59 57 16	60 00 36	60 03 20	60 05 27							
+4°					59 30 49	59 23 09	59 14 51	59 05 53	58 56 15	58 45 55	58 34 54
	59 55 09	59 59 42	60 03 38	60 06 58	60 09 42						
+5°						59 14 15	59 04 41	58 54 27	58 43 32	58 31 55	58 19 35
	59 52 25	59 58 11	60 03 20	60 07 53	60 11 50	60 15 10					
+6°							58 53 51	58 42 20	58 30 08	58 17 12	58 03 33
	59 49 03	59 56 03	60 02 25	60 08 11	60 13 21	60 17 55	60 21 53				
+7°								58 29 32	58 16 01	58 01 46	57 46 47
	59 45 05	59 53 18	60 00 54	60 07 53	60 14 17	60 20 04	60 25 15	60 29 51			
+8°									58 01 11	57 45 36	57 29 15
	59 40 28	59 49 55	59 58 45	60 06 59	60 14 36	60 21 36	60 28 01	60 33 51	60 39 05		
+9°										57 28 40	57 10 56
	59 35 12	59 45 55	59 55 59	60 05 27	60 14 18	60 22 33	60 30 12	60 37 15	60 43 43	60 49 35	
+10°											56 51 48
	59 29 18	59 41 16	59 52 35	60 03 18	60 13 23	60 22 53	60 31 46	60 40 03	60 47 45	60 54 52	61 01 24
h_A	0°	+1°	+2°	+3°	+4°	+5°	+6°	+7°	+8°	+9°	+10°

Fig. 91 The true horizontal values for the angle $P\hat{O}Q$ when the observed angle $A\hat{O}B = 60° 00' 00''$, for vertical angles $A\hat{O}P$ and $B\hat{O}Q$ between +10° and −10°.

For geographical work offshore, therefore, the sextant is excellent. At any desired point, the horizontal angles to previously established marks at or near sea level can be measured and recorded, and the angles $A\hat{O}B$ and $B\hat{O}C$ can be laid off on the chart, with a station pointer if available, or with protractor and tracing paper, to fix the position, always provided that A, B, C and O do not lie on or near a danger circle (see p. 92).

The *station pointer* used in conjunction with the sextant consists of a graduated circle with three movable radial arms. These can be clamped in any desired positions which give appropriate angles between them, and the instrument is then moved on the chart until the three arms pass through the plotted positions of the three fixed points observed. The centre of the circle then marks the observer's position. This is identical in principle with the method of drawing three rays on tracing paper in the case of plane table resection (see p. 91), and similarly, cannot be operated if the three fixed points and the observer's position lie on, or close to, a common circle.

It is clear that one cannot make simple use of the sextant for the observation of azimuth, owing the impossibility of simultaneous observation of both horizontal and vertical angles.

The Abney clinometer

The Abney clinometer is of very simple design. It consists of the following parts:

(a) A small metal tube of rectangular cross-section
(b) A telescopic eyepiece fitted into one end of this. The instrument can be used with the eyepiece extended or not as the individual observer prefers
(c) A smaller rectangular tube fitted into the other end of the main tube. The inserted tube has a silvered mirror, inclined at 45°, set at the inner end, but extending only half way across. A thin horizontal crosswire is set inside the inserted tube to act as a sight vane
(d) A semi-circular scale, graduated in degrees, is attached to the side of the main tube. Running against this scale is a vernier scale graduated

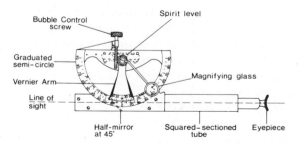

Fig. 92   The component parts of the Abney clinometer.

to read to 10′. A small magnifying glass aids reading. The vernier is on a radial arm pivoted at the centre of the semi-circle, and moved by means of a milled screw, allowing fine adjustment of position

(e) A small spirit level is attached to a bracket at the pivot of the vernier arm, but on the opposite side of the semi-circular scale. The bracket moves in the same plane as the vernier arm, but is at right-angles to it: it moves with the arm when the milled screw is rotated. The spirit level is attached to the bracket by two capstan-headed screws.

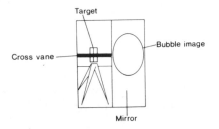

Fig. 93   View through the Abney clinometer. The left side is open apart from the sighting crossvane. The right side is the angled mirror. When the spirit level is horizontal the image of the bubble is visible and can be brought opposite the sighting crossvane.

There is an aperture in the top of the main tube immediately above the 45° mirror. The design ensures that the line of sight through the tube, when reflected by the mirror, passes through the centre of the spirit level, which has glass above and below to allow such viewing. If the small bubble is central in the spirit level its image can be seen in the mirror.

The Abney clinometer is used for the measurement of slope angles. The observer sights through the tube, bringing the cross vane on to the point observed. If the point is level with the instrument, so that the line of sight is horizontal, then the bubble will appear in the mirror alongside the crossvane, as in fig. 93; in this case the vernier arm is at zero and the axis of the spirit level is horizontal. If the slope of the ground is being measured, the object sighted must obviously be viewed at the same height above the ground as the observer's eye, so that the line of sight exactly parallels the ground surface. This is best achieved by holding the instrument at the 1.5 metre mark on a ranging rod and sighting to the corresponding mark on a pole held at the point required. With practice, the observer learns to judge heights above ground level when taking slope readings for correction of measured distances to horizontal distances, for example, so that ranging rods are unnecessary. The observer must guard against sighting to the eyes of a person of markedly different height from himself.

If the line of sight is inclined, as is usually the case, the image of the bubble may be brought back to the sight vane by turning the milled screw. The reflected line of sight always passes through the centre of the spirit level irrespective of the bubble position, and so as the rotation of the screw

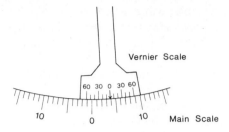

Vernier Scale

60 30 0 30 60

10   0   10   Main Scale

Fig. 94   The double vernier of an Abney clinometer. If the zero of the vernier moves to the right of the main scale zero, use the vernier to the right; and vice versa. In this case the vernier zero has passed 03° on the main scale but has not reached 04° One looks along the right hand vernier scale for a graduation which coincides with a main scale graduation, and the 50 is seen to be the only mark which meets this condition. The reading is therefore 03° 50′.

moves the spirit level to the horizontal the bubble returns to the observer's view. When the sight vane lies on the point observed, with the bubble intersected by the extended line of the vane, the angle of elevation or depression can be read from the circular scale and the vernier. The whole degrees are read from the position of the index arrow, and the fraction of the degree, to the nearest 10′ of arc, is read from the vernier. The angle can be estimated to the nearest 5′ of arc if desired. The vernier is double, so that it may be read in either direction from the zero index (see fig. 94). If the index arrow moves towards the observer, he must use the vernier that is on his side of the index. In this case it is an angle of elevation. If the index moves away from the observer, he uses the vernier on the further side of the index arrow, to read an angle of depression.

Figure 95 indicates the simple geometry of the instrument. The angle required is $H\hat{O}A = \theta$. The triangles OPX and XVM are similar, as the angles at X are equal and the angles at P and M are both 90°. Therefore $X\hat{V}M = P\hat{O}X = H\hat{O}A = \theta$. The angle $X\hat{V}M$ is measured directly from the instrument scale, as VM is the zero line of the scale.

Adjustment

It is obvious that to be in adjustment the instrument must have the bubble in the centre of the spirit level when the index arrow is at zero on the main

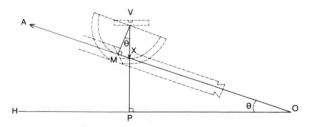

Fig. 95   The simple geometry of the Abney clinometer. The angle required is $P\hat{O}A = \theta = X\hat{V}M$, which is measured by the vernier arm VX.

scale and the line of sight is horizontal. An error will exist when the axis of the spirit level is not exactly at right angles to the vernier arm. The axis of the spirit level is controlled by the two capstan-headed screws fastening the level to its bracket.

To ensure that the Abney clinometer is in adjustment, proceed as follows. Sight to a well-defined distant object, and read the angle of elevation or depression. Let this angle be called θ. If the instrument is out of adjustment, the vernier index, when the bubble is central, will be to one side of the zero mark by an angle of error α. Therefore the reading taken to the distant object will not be the true angle θ but will be an angle $(\theta + \alpha)$. If the instrument is then inverted, a second reading can be taken to the object, once the bubble has been brought to centre. If the first reading showed an angle of elevation, with the index towards the observer, the inverted second reading will show a reading with the index away from the observer, and vice versa. As the index arrow is on the opposite side of the zero graduation on the semi-circular scale, the angle recorded will be $(\theta - \alpha)$. The mean of the two readings is θ. The vernier can then be set to read θ, and the capstan-headed screws adjusted, by tightening at one end of the level and slackening at the other, until the bubble is in centre when the sight vane lies on the object as viewed through the tube.

The aneroid barometer

The aneroid barometer consists of a circular metal case with a graduated face over which a pointer rotates. The graduated scale is in metres (or feet) with a second scale of pressure (millibars or inches) giving approximate correspondence for average conditions of the atmosphere. There is usually a vernier scale running outside the height scale, and moved by rotation of a milled knob at the top of the instrument. To take a reading, the vernier is turned until its zero comes immediately beneath the pointer. The main scale divisions are read from the pointer, and the vernier is used to read subordinate intervals. Care must be taken in the reading, because the pointer is inevitably some little distance above the graduated scale, and a parallax error can be introduced if the observer does not view the pointer perpendicularly. A magnifying glass on a circular ring assists reading at all points on the scale.

The instrument case contains a thin metal box which has been exhausted of air. This vacuum box will respond to quite small changes in atmospheric pressure. One side of the box is attached to the instrument base, the other is free to move. Changes in atmospheric pressure therefore alter the volume of the box, and alter the position of the box top relative to the base. This movement is magnified by a system of delicate levers within the case, and is ultimately translated into rotational movement of the pointer across the graduated scale. The system is very carefully pivoted throughout in order to introduce the minimum frictional resistance and is further balanced by

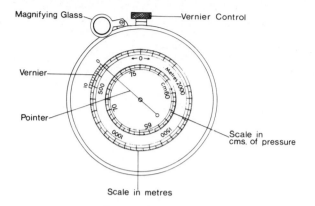

Magnifying Glass — Vernier Control

Vernier —

Pointer —

Scale in cms. of pressure

Scale in metres

Fig. 96   An aneroid barometer.
   The vernier control knob moves the vernier to a position beneath the pointer so that the latter is aligned with the vernier zero. The instrument is graduated in centimetres of mercury atmospheric pressure with an approximately equal altitude scale in metres.

fine springs. Any jolting of the aneroid may therefore disturb the system and may, in particular, shake one or other lever off its pivot. If this happens, the springs generally cause the pointer to move to the maximum or minimum reading possible and to remain immovable at this value until repaired by a competent mechanic. It is not advisable to attempt the repair of a damaged aneroid without professional skill and equipment. A common cause of jolting is failure to fasten securely the leather carrying case. Many patterns of case are inadequately attached to the carrying strap so that they are not balanced when the instrument is in the case. This can lead to sudden inversion of the case and the falling-out of the instrument if the fastening strap has not been secured. A second cause of trouble is over-energetic tapping of the instrument before reading. There is often a slight delay in the response of the pointer, due to friction in the lever system. This is known as hysteresis, and can be overcome by a light tapping of the aneroid with a finger to ease the pointer through the last small movement. "Light tapping" means at most one or two very gentle taps with the end of one finger nail: anything more violent may shake loose the lever system.
   The only adjustment allowable to the observer is the setting of the pointer, achieved by the very slow rotation of a small screw sunk into the back of the case. It is advisable to set the pointer to read considerably above the true height (see p. 112).
   Aneroids can be obtained to cover any range of altitudes required. For most geographical purposes, the range 0–1000 m is adequate, but a range up to 10,000 m is available and many mountain or continental plateau areas will require a range up to 3000 m. An aircraft altimeter is a very similar instrument, indicating approximate height above sea level, albeit with much greater range in altitude, but with an additional control allowing movement

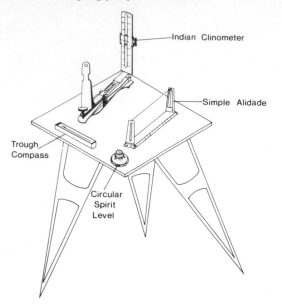

Fig. 97   The plane table on its tripod, with accessories.

of a scale to read zero at the altitude of the landing ground next being
used, so that height above ground can be read directly from the pointer.

The plane table

The plane table is a wooden board, usually about 0.6 x 0.75 m, or 0.6 x
0.45 m, fixed to a tripod by a wing-nut screw on its underside. If the screw
is only partially fastened, the board is held to the tripod but is free to turn
in a horizontal plane. If tightening is completed, the board is clamped in a
fixed position. The board is carried in a flat canvas case.

The board has a very smooth upper surface, and great care should be
taken to ensure that this is never marked or scarred. The tripod should be
of a substantial nature with broad-topped frame legs giving a firm foundation. Telescopic tripods are easier to carry, but inevitably destroy much of
the accuracy of the work, and telescopic tripods should therefore be avoided.
The wing-nuts holding the legs to the tripod head should be tightened
gently at each station before work starts, to eliminate "shake" in the tripod.

The paper on which the map is drawn is mounted on the board. Figure
98 shows the method of mounting. The paper is shaped to the underside
frame of the plane table, and is pinned on the underside. Pins must NEVER
be used on the upper surface of the plane table, as pinholes in the surface
of the wood wreck the drawing surface and can lead to tearing of the map
if the pencil point drops through the paper into a pin hole. If time and
circumstances permit, the paper, after shaping, is soaked in a bath of water

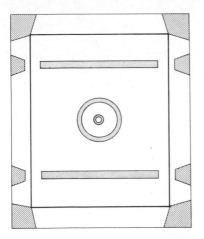

Fig. 98 To mount a plane table, lay the table, upper surface downward as shown, on a sheet of paper larger than the table. Pieces (as shown by shading) are then cut from the paper, so that when the paper is folded back for fastening on the underside of the table, the folded parts will not coincide with the metal fittings and will lie flat against the wood

and is then laid face downwards on a sheet of clean glass and all air bubbles are pressed out of it. The board is then in its turn laid face downwards on the paper, the flaps of the latter are folded over to the back of the table, as tautly as is possible without tearing the wet paper, and are pinned down. The paper, stretched when wet, shrinks again on drying, and should fit exactly to the table top with no air bubbles between board and paper. If there is no opportunity to mount the board in this way, and the paper is applied dry, it should be stretched as tightly as possible before folding over the edges of the board. If air bubbles remain after mounting, the paper should be unfastened and the operation repeated, as air bubbles beneath the paper lead to inaccuracies in line drawing. If the sheet is being marked with coordinated fixed points *before* mounting, only dry mounting is possible, as the degree of stretch following wet mounting cannot be exactly foreseen, possibly leading to different amounts of distortion in different directions, and contraction on drying is prevented by the pins from equalling expansion on wetting, so that scale distortion is inevitable. If the work is going to continue for more than 2 or 3 days, it is preferable to paste down the edges of the paper on the underside (or to fasten the edges with adhesive tape) as with pins the paper slowly works loose by enlargement of the pinholes.

Accessories with the plane table

Alidade The plain alidade is a wooden straight-edge with a slit back sight and a hair-wire foresight in a metal frame. The sights are so placed that a line joining them will be exactly parallel to the boxwood edge. Some alidades are made of metal instead of wood, and the better models have a parallel rule attached to the base plate and capable of being swung out from it.

When drawing a ray to another point, the observer aligns the back sight and foresight with the object, and moves the parallel rule sideways so that a line drawn along its edge, with the pencil held vertically, will pass through the observer's plotted position. If there is no parallel rule, the alidade must be pivoted around a finger tip or the butt end of a pencil on the observer's plotted point: it must not be pivoted around a pin or a pencil point which can damage the map and the surface of the table. It is clear that the foresight wire must be completely taut in its frame, and a broken wire can be replaced by pulling a hair from the head and using matchstick plugs to get it taut in the frame, or it can be used with the tightening screws holding the wire in some patterns of frame. It is useful to have a thin cord attached between the tops of the back sight and foresight, so that the cord is taut with the frames erect. If the line of sight is too steep for the target object to be visible in the normal way, it can be sighted using back sight and cord, or cord and foresight (depending on whether it is an angle of elevation or depression), as the back sight slit, the cord and the foresight vane all lie in the same vertical plane.

Some alidades have a telescope instead of open sights. This allows more precise alignment and also permits sighting over longer lines. A telescopic alidade has a vertical circle, in whole or in part, allowing measurement of vertical angles also. Controls on the telescopic alidade vary with manufacturer, but every instrument should have an eyepiece (rotation of which clarifies the crosswires in the field of view), a focussing screw or ring, a telescope clamp screw and a vertical slow-motion tangent screw. There will be a vernier or a micrometer for reading the vertical circle, and one or more control screws for levelling the instrument lengthwise and crosswise, with spirit levels for this purpose. Some instruments have additional refinements (see p. 86 for the telescopic alidade fitted with Beaman arc). The telescopic alidade usually has stadia lines so that with telescope horizontal it can be used for tacheometric measurement of distance, and with telescope inclined the true horizontal distance can be calculated from the stadia readings by the use of Table 4A. There is one possible source of error in that with the spirit level bubble central and the vertical circle reading zero, the

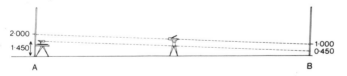

Fig. 99   Adjustment of the telescopic alidade.
    The difference in height between *A* and *B* is determined with a level sited midway between the two graduated staves.
    The plane table is then placed at *A* and the height of the telescope axis recorded. The reading to be expected at *B* is then calculated, and with the vertical circle at zero the telescope is brought to read the figure (in this case 0.450 m) by means of the alidade levelling screw. The spirit level locking nuts are then adjusted to bring the bubble to centre.

line of sight may not be horizontal. This needs adjustment for collimation in altitude, and follows the same principles as the corresponding adjustment of the level or the theodolite. The simplest method of adjustment is as follows:

(1) Set a level (known to be in adjustment) midway between points A and B, of roughly equal altitude, each about 50 metres from the level. With the telescope horizontal, take staff readings to A and B, and calculate the difference in altitude of B relative to A.

(2) Establish the plane table at A, with the telescopic alidade on the board. Measure the height of the telescope above the ground. Calculate the reading to be expected on the graduated staff at B if the line of sight is horizontal. Bring the horizontal crosswire on to this graduation, keeping the vertical circle reading zero, by use of the levelling screw of the alidade. Then adjust the screws holding the spirit level until the bubble is central.

Indian clinometer   The Indian clinometer is illustrated in fig. 100. Two vertical standards hinge up from the ends of a brass frame, which houses a small spirit level. The back sight has a small circular peep-hole; the foresight has a wide central space between the graduated edges of a metal frame. A slide with a horizontal crosswire is racked up and down the frame by means of a milled screw. Both sides of the frame have a zero mark at the same height above the base as the back sight hole. One side of the frame is graduated in degrees and fractions, and the other is graduated in the corresponding natural tangents. Graduations increase away from the zero mark upward for elevations and downward for depressions. The instrument as described rests on a second metal base plate, to which it is hinged at its forward end. The near end has a levelling screw threaded through the brass frame, and bearing on the base plate: rotation of the screw raises or lowers the rear end of the instrument and allows the bubble of the spirit level to be centred.

In use, the clinometer stands on the plane table and is aligned with the object being observed. The bubble is brought to centre with the levelling screw and the foresight wire is brought on to line between the back sight and the object. The tangent value is then read from the graduated standard. The difference in altitude between the back sight and the object is $s \tan \theta$ where s is the true horizontal distance and θ is the angle of elevation or depression. If the object is already plotted, as well as the observer's position, s can be scaled from the map. Tan θ is read directly. If the altitude of the object is known, the altitude of the clinometer back sight can be calculated; subtract the height of the back sight above ground level and the altitude of the latter becomes known. Alternatively, if the observer's altitude is known, the altitude of the object can be calculated. Calculation is simplified if the foresight is brought on to the object at a height above ground equal to the height above ground of the clinometer back sight.

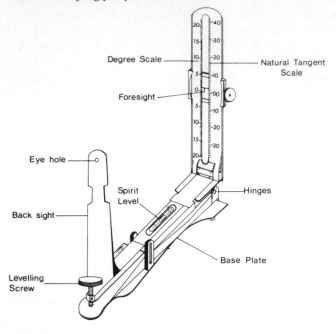

Fig. 100   The Indian clinometer.

The levelling screw raises or lowers the end of the plate carrying the back sight and the spirit level. The bubble is brought to centre, the object is sighted through the back sight eye hole and the foresight is racked down until its crosswire is aligned with the object being investigated. The angle of elevation or depression can then be read from the scale of degrees, or the corresponding natural tangent can be read from the vertical scale.

To be in adjustment, the line of sight must be horizontal when the foresight is at zero and the bubble is central. The adjustment is very similar to that of the telescopic alidade, and can be done at the same time. Knowing the altitudes of points *A* and *B* in fig. 99, and the height of the back sight above ground level, it is easy to calculate the reading that should be obtained on a staff at *B* if the error is nil. The foresight is set at zero, and the observer, looking through the back sight, brings the crosswire on to the desired graduation by turning the levelling screw; the bubble is then returned to centre by turning its holding screws, usually capstan-headed. If the staff is too distant for graduations to be read with ease, a piece of white tape is put across the staff at the required reading and the crosswire is brought on to this.

Trough compass   A simple compass needle is mounted in a rectangular case with a glass top. At the ends of the needle are scales graduated to 5° each side of zero, so that the box can be turned until the needle reads zero at each end. A line drawn along the edge of the box then defines the magnetic meridian.

A small stud on top of the box at one end of the glass rests on a bent metal strip, so that if the stud is pressed the far end of the strip lifts the compass off its pivot. The stud is automatically depressed when the wooden lid of the box is slid into place, so that the pivot cannot be damaged in transit.

Circular spirit level   This is a very simple piece of equipment used for levelling the plane table. It is placed on the table when a new station is reached, and the bubble is brought to centre by moving the tripod legs. In general it is best to get two legs firmly set in the ground, and then to move the third leg sideways until the bubble lies on a line between the third leg and a point midway between the other two legs. The bubble can then be brought to centre by moving the third leg on this line either inwards or outwards.

Plumb bob   For small scale work the table can be centred by eye to within 50 mm of a station mark, and if this is less than the plottable error the centring is adequate. For large scale work more precise centring is necessary, and this is particularly true of plane table traverse work. For accurate centring a plumb bob should be suspended from the winged clamping nut on the underside of the tripod head. On occasion it may be necessary to hold the plumb bob vertically beneath a particular point on the map and to fix the table with this point plumbed over the station mark.

Scales   For all measurements on the map a boxwood scale is required with graduations in millimetres (or, if feet are being used, in inches and fiftieths). A scale appropriate to that being used on the map, e.g. 1:5000, can be employed if preferred.

Drawing equipment   Hard pencils, soft erasers and a razor blade are desirable in the field.

Linen tape   A tape is required for short horizontal measurements and for height measurements to instruments. A linen or plastic tape of 30 to 50 m, wound in a circular case, is necessary. The tape is usually graduated in metres and centimetres on one side and in feet and inches (or feet and tenths) on the other. The zero mark is usually at the edge of the brass ring at the end of the tape. A hinged handle for rewinding into the case is set into one side of the case. The tape should always be cleaned before winding and should be wound through two fingers held over the case aperture to ensure that there is no twisting of the tape on the spool.

The prismatic compass

Figure 101 illustrates the components of the prismatic compass.

The compass needle is fastened to the under side of a circular, non-magnetic disc or ring, graduated on its outer edge in degrees and half-degrees.

The needle is mounted on a pivot in the centre of a shallow, circular brass case. A glass plate, held by a metal ring, covers the top of the case.

At one edge of the case is a hinged frame carrying a vertical hair or wire held at top and bottom by a small wooden plug. This is the foresight. It is essential that the wire be kept taut, and this is done by pressing home the plugs if necessary. A human hair is an adequate replacement for a broken

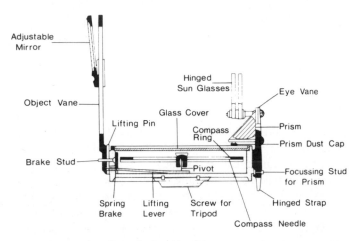

Fig. 101   Section through the prismatic compass (see text).

wire. This foresight frame, when not in use, folds down to lie on the glass cover of the case. In this position it presses on a small brass bar near the hinge, the other end of the bar resting on the raised end of a bent metal strip on the floor of the case. Pressure on the outer end of the strip forces this down, and raises the inner end, which lifts the needle off the pivot and so avoids damage during transport.

Diametrically opposite the foresight frame is a magnifying prism enclosed in a brass case. The latter is slotted to act as a back sight. When not in use this prism hinges back and is held against the outer edge of the case by a small folding bracket. The under side of the prism has a small hole in the case, in the operational position pointing vertically downwards to the graduated disc. A small dust cover protects the prism when not in use, and must be swung sideways before a reading is attempted. Small tinted glass discs are also attached to the prism case, and can be swung into the line of sight if a bearing is being taken to a luminous signal or to the sun.

On the outside of the case, at the base of the foresight, is a small brass stud. The inner end of this lies against a spring strip inside the case. Pressure on the stud forces the strip inwards until it touches the graduated disc and checks any movement of this. It is used to dampen oscillations of the needle and so to save time: if the needle is swinging over a wide arc, a check is made with the stud approximately in the centre of the arc, after which the stud is released and the needle allowed to complete its alignment on the meridian.

A screw socket on the under side of the case allows the compass to be supported on a tripod, usually of light weight with telescopic legs and a ball-and-socket joint at the top. This is fastened by a ring when the compass is approximately horizontal. The compass can be used without the tripod, being held in the hand, preferably with thumb and first finger gripping the folded-down bracket which secures the prism when travelling, and with the case resting on the other fingers.

A reading is taken by alignment of the target object, the foresight wire and the back sight slot. When the needle has come to rest the bearing can be read, the image of the graduated circle being visible in the prism while the sights are kept on line. If the figures are not clear they can be focussed by a small vertical movement of the prism, which has about 5 mm of available slide. It is essential that the case is horizontal, as otherwise the graduated ring may touch the floor of the case on one side and may cease to swing. Note that the prism reads from the observer's side of the circle, instead of from the foresight end. All circle readings are therefore graduated to give the correct answer as seen in the prism: if viewed without the prism they appear to be in error by 180°. For example, if the observer sighted northwards along the magnetic meridian, the North end of the needle would be at the foresight and the South end of the needle under the prism, but the prism reads 0°, this being the graduation beneath it at the South end of the circle. This is mentioned in order to avoid possible error if for any reason the prism is unusable (as in heavy rain, if water enters the prism case) and an attempt is made to read graduations by looking directly at the circle through the glass cover.

The foresight has sliding on it a small hinged mirror. The purpose of this is to permit bearings to be taken over steep slopes. Figure 102 indicates the lines of sight for steep upward and downward lines. Note that in the case of a steep upward slope two methods are possible. Either one views the reflected

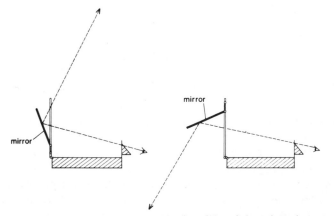

Fig. 102   Use of the mirror on the foresight of the prismatic compass for bearings to points that would otherwise be above or below the line of sight.

and inverted image, facing forward, or one views the reflected image by standing with one's back to the forward station. In the latter case the bearing read will need to be corrected by 180°.

Some patterns of compass have an oil-filled case instead of allowing the needle to swing in air. The oil damps down oscillations and makes reading quicker. Against this is the disadvantage that if an air bubble enters the case it tends to press on one side of the circle and to upset readings accordingly.

Bearings

Magnetic bearings are measured from the magnetic meridian in a clockwise direction from 0° to 360°. Magnetic North and True North rarely coincide: at most places there is an angle between the two, known as *Magnetic Declination.* Magnetic Declination may be East or West of True North. An *isogonic line* passes through places all having the same Magnetic Declination.

As the positions of the earth's magnetic poles vary from year to year, the alignment of the magnetic meridians also varies, and so, also, does the magnetic declination at all points on the earth. The annual rate of change at any point is the *Annual Magnetic Variation*, which indicates the rate and direction of change. See p. 97 for the calculation of up-to-date Magnetic Declination values, and for comments on bearings.

Battery-operated calculators

There are now available small pocket-sized calculators such as the H.P. 35 which include keys for logarithms and for trigonometrical functions. These greatly facilitate survey calculations. Some of their advantages are given below.

Laboratory work

Logarithmic and trigonometrical tables are unnecessary, and it is, for example, possible to calculate a theodolite traverse directly for differences of Eastings and Northings. Each bearing is converted to decimal notation, is entered in the instrument store register and is then recalled in turn for sine and cosine, being multiplied directly by the horizontal distance to obtain the Eastings and Northings movement over the line in question.

Conversion of angles or bearings to decimal notation is achieved very quickly by entering the number of seconds, dividing by 60, adding the minutes, dividing by 60, and adding the degrees. If one is calculating azimuths, for example, an opposite routine allows a tangent value to identify the appropriate angle in degrees, minutes and seconds. The operator should familiarise himself with trigonometrical functions from tables before using such calculators, and + and − signs are very important in the angular calculations.

Field work

Tacheometry can be done directly without recourse to the tables, provided that the correct formulae are used, and most survey operations can be accelerated. The great advantage is that many calculations can be completed and checked before leaving the field, so that errors can be identified and eradicated immediately, with a possible time saving of whole days.

The latest models include some which can be programmed automatically for specific surveying calculations. It should be noted that the H.P. 65, for example, is programmed for use on American practice for interpolation in tables of solar declination, giving slightly different answers than interpolation from 2-hour British tables, although the error, which will be more or less consistent over a set of observations, is unlikely to exceed about 30" in azimuth. There are other small differences between U.K. and U.S. practice which affect professional use of the programmes but which can be ignored by the field scientist.

Radio telephones

"Walkie-talkie" units greatly simplify survey operations.

Control networks

In laying out a control network, it is advantageous if four individuals, each equipped with a receiver/transmitter, co-operate in the location of points. They can check for mutual intervisibility, for the well-conditioned nature of angles at each point, and for suitability of each point for the mapping of local detail. Shift of location of any point is thus made as part of the whole figure without the necessity for retracing steps to other stations to check on intervisibility.

Detail fixation

The mapping of detail benefits enormously, as the range of workable sights is increased. If stadia readings are being taken for distance measurement, the target staff can be worked comfortably at ranges of 200 m without recourse to shouting; each location can be identified and reported by the staff holder to the plane table operator, e.g. "I am at the point where the footpath passes from heather to bracken", "There is a small tributary joining the main stream here", "This is the inner edge of the terrace at the base of the steeper slope", etc. Exchange of such information reduces the necessity for record-keeping by both parties for checking later.

It is also possible to extend the range of observation to about 400 m. If the observer cannot identify a staff graduation with complete clarity at long range, he can nevertheless direct the staff holder's attention to the relevant mark (by asking him to slide a finger or other object to the mark) and can obtain the detailed reading by radio from the staff holder.

Aneroid readings

It is obvious that accuracy is improved if all aneroid readings with the moving instrument are matched by simultaneous recordings on the base instrument, if two operators are working together.

Specialised readings

Surveys are usually combined with other scientific observations, and inter-communication by radio may be invaluable. Examples are simultaneous microclimatological readings across a network of surveyed points, the timing of floats in geomorphological studies of streams or offshore movement, the notification of movement of small animals in any area of zoological study, the tracking of individual vehicle routes in an urban traffic study, etc.

Other instruments mentioned in the text

The Ewing Stadi-altimeter is described in chapter 2 (see pp. 73–5).

5 Simple astronomy

To obtain values for azimuth, time and latitude, use is made of the astronomical triangle.

Data for the sun and other heavenly bodies are catalogued in almanacs published annually. The location of a heavenly body can be stated in one of three ways:

(a) Declination and Right Ascension
(b) Declination and Hour Angle
(c) Altitude and Azimuth.

Observations are made on the third system, but data are tabulated on the first system, this being the only one of the three which is independent of the observer's position.

Declination and Right Ascension

In fig. 103, P_1 and P_2 are the poles of the celestial sphere on which it is assumed that all bodies outside the solar system are fixed. The solar system bodies are assumed to lie on the sphere but to change their positions at an appreciable rate. QR represents the equatorial plane produced to intersect the sphere, just as P_1 and P_2 lie on the extended axis of the earth. The plane of the equator is the main plane of reference. To an observer on the earth it appears that the sphere rotates on its polar axis once every 24 hours, and S_1 and S_2 are stars with the directions of apparent rotation shown. Through S_1 and S_2 are drawn declination circles, i.e. great circles passing through the

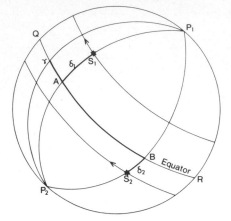

S_1A = North declination of S_1 = δ_1
S_2B = South declination of S_2 = δ_2
ΥA = Right Ascension of S_1
ΥB = Right Ascension of S_2

Fig. 103   Declination and Right Ascension.
   The diagram represents the hypothetical celestial sphere on which are located the heavenly bodies. The earth is assumed to be at the centre of the sphere.
   The star S_1 can be located firstly by its *declination*, S_1A in the figure, which corresponds to latitude on the earth in that it can be defined by an angular distance North or South of the equatorial plane. Longitude is replaced by *Right Ascension*, which is the equatorial distance measured from the celestial meridian through the First Point of Aries Υ, i.e. the angular distance ΥA in the case of S_1.
   The celestial sphere appears to rotate once round the earth every 24 hours, as shown, so that δ and R.A. have no relationship to the observer's position. As Υ and the stars are all on the sphere their relative positions do not alter appreciably, and δ and R.A. are virtually constant for a star. The sun and other bodies in the solar system vary in δ and R.A.

poles and through the star positions. The *declination* (δ) of a star is the arc of its declination circle intercepted between the star and the equator. The declination is regarded as positive if the body is north of the equator and negative if south of the equator.
   The *North Polar Distance* is the arc of the declination circle between the body and the North Celestial Pole. The South Polar Distance is measured from the body to the South Celestial Pole. The Polar Distance is usually, but not always, equal to ($90° - \delta$); on occasion an observer in one hemisphere may need to use a body in the other hemisphere, so that, for example, he may require the North Polar Distance of the sun in November, when the sun has a South declination, and the North Polar Distance is then $90° - (- S.\delta) = 90° + S.\delta$.
   The declination of a body thus corresponds to "latitude". "Longitude" is measured at right angles to declination, i.e. as an arc of the celestial equator. The reference point on the equator is the intersection with the declination circle of the 1st Point of Aries (Υ). The 1st Point of Aries is that point on the heavenly sphere which would be intersected by a line

drawn from the earth's position at the vernal equinox through the centre of the sun. (This point alters slowly in location due to precession of the equinoxes, the sun arriving at the position of the vernal equinox a little earlier each year, but the movement is not large enough to invalidate the 1st Point of Aries as a reference mark.)

The *Right Ascension* of a heavenly body is the arc of the celestial equator between the declination circle of the body and the declination circle of the 1st Point of Aries. It is reckoned eastward from the 1st Point of Aries, i.e. in the opposite direction to the apparent rotation of the heavenly sphere, and is measured in time from 0 to 24 hours. In Figure 103 ΥA and ΥB are the respective Right Ascensions of the stars S_1 and S_2.

Declination and Hour Angle

In this system of definition, the declination of the heavenly body is as above. The second plane of reference is not the declination circle of Υ, but the meridian of the observer; from this is measured the *Hour Angle*, which is the arc of the equator between the observer's meridian and the declination circle

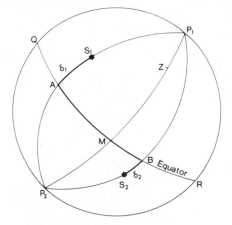

S_1A = North declination of S_1 = δ_1
S_2B = South declination of S_2 = δ_2
P_1ZMP_2 = declination circle through observer's zenith
AM = Hour Angle of S_1 (+ve)
MB = Hour Angle of S_2 (–ve)

Fig. 104   Declination and Hour Angle.

Declination of the star S_1 is measured as in fig. 103. Equatorial distance is measured not from the meridian of the First Point of Aries but from the projected meridian of the observer on earth. This distance is defined in angular terms as the *Hour Angle*.

As the celestial sphere rotates around the earth, the declination of a star S_1 remains unaltered but its Hour Angle will be constantly changing, being zero when the star is on the observer's meridian (i.e. when S_1 is between Z and M). The Hour Angle is given in terms of time, so that in this diagram S_1 has a positive Hour Angle, which represents the time in hours, minutes and seconds since the star crossed the meridian PZM. Star S_2 has a negative Hour Angle, which is the interval that has still to elapse before it will reach the meridian.

of the body. It is measured westward from the observer, i.e. in the same direction as the apparent rotation of the heavenly sphere, and is again given in units of time from 0 to 24 hours. However, it is customary to quote the Hour Angle as less than 12 hours for computation purposes, calling it + if measured westward from the observer and − if measured eastward.

Whereas Declination and Right Ascension are both constant values which can be tabulated for all users, the Hour Angle is obviously not only different for observers at different points, but varies continuously at any one point. When the star transits the observer's meridian, the Hour Angle is 0 hours (or 12 hours at lower transit below the horizon). Subsequently the Hour Angle increases as the star moves away. A positive Hour Angle is the interval of time since the star crossed the observer's meridian, or the negative Hour Angle is the interval before transit will take place.

In fig. 104 Z is the zenith, the point on the celestial sphere which is overhead at the observer's position. The great circle $P_1 ZMP_2$ is the declination circle of the observer's meridian. As before, ΥA and ΥB are the respective declinations of the two stars S_1 and S_2. The corresponding Hour Angles

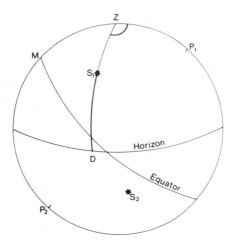

$S_1 D$ = altitude = h
$P\hat{Z}S_1$ = azimuth of S_1
S_2 is below the horizon

Fig. 105   Altitude and Azimuth.

The observer on the earth locates a heavenly object by its *altitude*, or angular distance above the horizon, and its *azimuth*, or bearing from the observer. In this case S_1 has the altitude $S_1 D$ (fig. 108 shows that the angle between the plane of the horizon and the plane of the equator = 90° − latitude). The azimuth is, like all bearings, measured clockwise from North. In this case the observer's meridian is in the plane of the paper, Z marking the zenith or point on the celestial sphere above the observer's head. If the observer faced North he would look along the meridian plane in the direction ZP_1 (P_1 being the elevated pole here assumed to be North) and would turn through the azimuth angle $P\hat{Z}S_1$ in order to look towards the star S_1. As shown, S_1 has an azimuth of about 100°, i.e. it is a little south of east when viewed by the observer.

are MA (positive) and MB (negative) given as arcs: these can be defined in angular form by the angles $ZP_1 S_1$ and $S_2 P_1 Z$.

Altitude and azimuth

Both altitude and azimuth are temporary qualities peculiar to the observer's position, and lasting for only a moment of time. They are observed through the instrument telescope. The primary plane of reference is not the plane of the equator, but the plane of the observer's horizon. The *zenith* is the point on the heavenly sphere directly above the observer. A vertical circle such as that of the instrument on which altitudes are measured lies in the plane of a great circle passing through the zenith and at right angles to the plane of the horizon. The *celestial meridian* is a great circle whose plane passes through the zenith and both poles ($P_1 ZMP_2$ in figs. 104 and 105). The *Prime Vertical* is the great circle on the celestial sphere which passes through the zenith but is at right angles to the celestial meridian; it cuts the celestial equator in East and West points. The *altitude* (h) of the body is the arc of a vertical circle intercepted between the body and the horizon ($S_1 D$ in fig. 105). The *zenith distance* is the complement of the altitude, being the arc of the vertical circle between the body and the zenith, and therefore equals ($90° - h$).

The second plane of reference is again the observer's meridian. The *azimuth* is the horizontal angle between the plane of the meridian and the vertical plane passing through the body. It is measured clockwise from North and is given in degrees, between $0°$ and $360°$.

The astronomical triangle

Figure 106 shows superimposition of the relevant features of figs. 103, 104 and 105.

MZ is the observer's latitude ϕ (the angle between the plane of the equator and the vertical line through the observer)

ZP is therefore the co-latitude = ($90° - \phi$)

ZS is the zenith distance as defined above = ($90° - h$)

PS is the Polar Distance ($90° \pm \delta$), as SA is the declination of the body

ZPS is the Hour Angle (negative in the diagram, as S is moving towards upper transit of the observer's meridian MZP)

PZS is the azimuth angle, measured clockwise from the meridian ZP

PSZ is the parallactic angle and has no significance in the calculations.

Three sides and two angles of the triangle are therefore significant. If any three of these parts are known, the other two may be calculated.

The triangle is a spherical triangle. The geographer will usually know his latitude (ϕ), can measure the altitude of the body (h), and can read from almanac tables the declination of the body at the time of observa-

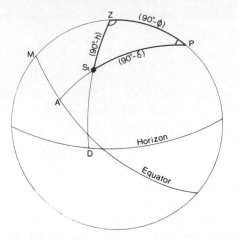

MZ = observer's latitude   = ϕ
AS_1 = North declination of S_1 = δ
DS_1 = Altitude of S_1 = h

$P\hat{Z}S$ = azimuth of S_1 at observer's position
S_1PZ = Hour Angle of S_1 at observer's position (−ve)

Fig. 106   The Astronomical Triangle.
     A combination of figs. 103, 104 and 105. From fig. 103, S_1A is the declination δ. From fig. 105, MZ is the latitude ϕ of the observer (since Horizon to Zenith = 90° and the angular distance Equator–Horizon = 90° − ϕ). Also from fig. 105, S_1D is the altitude b, and $P\hat{Z}S_1$ is the azimuth angle of S_1. From fig. 104, the angle between the plane of the observer's meridian and the declination circle of the star = the Hour Angle, in this case negative as the star is approaching the meridian.
     Five parts of the triangle ZPS_1 are therefore defined.

tion. He needs most commonly to discover the azimuth angle, which can be called A, and to check the observation times taken from his watch, i.e. to check the Hour Angle, which can be called t. The essential formulae are:

$$\tan \frac{t}{2} = \sqrt{\cos s \sin (s - b) \operatorname{cosec} (s - \phi) \sec (s - p)}$$

where

$$s = \frac{b + \phi + p}{2}$$

and

$$\tan \frac{A}{2} = \sqrt{\sec s \sin (s - b) \sin (s - \phi) \sec (s - p)}$$

In these formulae

    b = *altitude* above the horizon
    ϕ = *latitude* of the observer
    p = *polar distance* = 90 ± δ

and are listed in alphabetical order to assist memory.

Time

Accurate recording of time is necessary for astronomical work. It will be realized from the study of the astronomical triangle that changes in the Hour Angle will affect the other elements in the computation.

There are several different ways of measuring time. A *day* is measured as the interval between successive lower transits of a heavenly body across the observer's meridian. (Either upper or lower transit must be specified; if the declination of the body exceeds the latitude of the observer, both upper and lower transits will be above the horizon and the body will describe a circumpolar course in the sky. In this case the star passes between the zenith and the pole at upper transit, and crosses the meridian again on the other side of the pole at lower transit.) A *sidereal day* is the interval between successive lower transits of a star; a *solar day* is the interval between successive lower transits of the sun. *Local Sidereal Noon* is the moment of upper transit of the 1st Point of Aries across the observer's meridian.

Solar time

A solar day starts and ends with the lower transit of the sun across the meridian. Solar Noon is at upper transit, when *Local Apparent Time* (L.A.T.) = 12 hours. L.A.T. at any time is the Hour Angle of the sun at that instant measured westward from lower transit across the meridian of the observer, and is "sundial time".

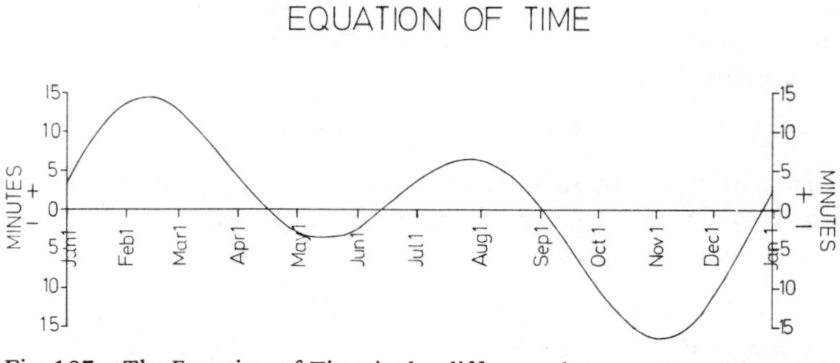

EQUATION OF TIME

Fig. 107   The Equation of Time is the difference between Mean Time and Apparent Time. The latter is based on observation of the real sun, and varies its rate during the year depending on the earth's position in its orbit. Mean Time is, in effect, Apparent Time averaged throughout the year to keep a constant rate. As clocks keep a constant rate, we have to work on Mean Time, calculated for a fictitious Mean Sun, while observations must of necessity be made to the real or Apparent Sun. In making calculations one has to use the formula

Local Mean Time = Local Apparent Time ± Equation of Time

The graph shows the variation in sign and magnitude of the Equation of Time throughout the year: this is the correction to be applied to sundial time in order to obtain Local Mean Time. (A further correction for longitude is necessary to correct L.M.T. to Greenwich Mean Time.)

The sun's apparent motion is not convenient for use, as its rate of movement varies during the year. This is because of the elliptical orbit of the earth around the sun, and the varying of rate of movement in the orbit, together with the inclination of the earth's axis to the plane of the orbit. It is therefore preferable to think of a *Mean Sun* which is supposed to move at a uniform rate and to have uniform intervals between passages over any meridian. This regular interval is a *Mean Solar Day* and is the mean length of true solar days throughout the year. The *Local Mean Time* (L.M.T.) at any instant is the Hour Angle of the Mean Sun at that instant, measured westward from lower transit across the observer's meridian.

Because calculations are based on the fictitious Mean Sun, while observations are necessarily taken to the Apparent Sun, it is necessary to know the variation of the former with respect to the latter. The relationship is given by the formula

L.M.T. = L.A.T. ± Equation of Time

The Equation of Time varies throughout the year in sign and magnitude. It is zero four times per annum, in mid-April, mid-June, at the beginning of September and on Christmas Day. Its variation and magnitude are shown in fig. 107: the Nautical Almanac gives daily values. Calculation of sun observations for time and azimuth involve adjustment for the Equation of Time.

Local Time values must, of course, be converted to Greenwich Time values as the watch used will record Greenwich Time ± an appropriate value for the observer's Time Zone. The variation of Local Time from Greenwich Time (whether L.A.T./G.A.T. or L.M.T./G.M.T.) is a function of longitude, at the rate of 1 hour per 15° longitude, or 4 minutes per degree. The correction formula is

G.M.T. − L.M.T. = West Longitude expressed in time

Sidereal Time differs from Solar Time, but is not discussed here. An observer using stars rather than the sun should be a specialist with knowledge of the necessary time conversion.

Latitude and longitude observations

These will normally be made only by the specialist. However, one simple method is given here in the belief that a geographer should be able to measure, at least approximately, his position on the surface of the earth. The non-specialist will, in general, wish to avoid the complications of star observations, and will find it easier to observe to the sun in daylight. Unfortunately the speed of sun movement, the changes in its declination and the necessity of observing to opposite limbs (with the impossibility of perfect balancing of paired observations) all combine to make sun observations relatively unsatisfactory for positional work.

The non-specialist can take approximate values for latitude from sun observations, if he wishes, but these will be far from precise. A modified form of the method of circum-meridian altitudes is based on repeated observations for altitude and azimuth over a period of time each side of Local Apparent Noon. These give altitude values which increase to L.A.N. and then decrease, and altitudes can be plotted against time on a graph. The highest point on the resulting smooth curve on the graph should be at the

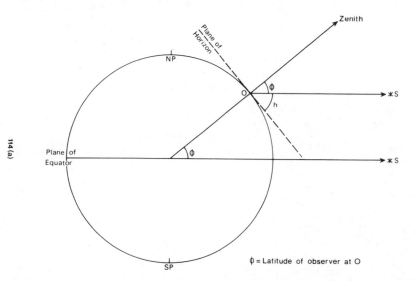

Fig. 108   At the equinox the sun is overhead at the equator (i.e. the sun has zero declination). At the observer's position O the altitude h of the sun is $(90° − \phi)$ where ϕ is the latitude.

moment of Local Apparent Noon, and at that moment the sun was due south (or north) of the observer. Interpolation between the horizontal circle readings of the observed solar positions either side of the highest point will give the circle reading for a point on the observer's meridian, and a mark can be laid out using the theodolite telescope. Azimuths can then be found by using the angle between this marked line and any desired points, remembering that if the marked line is south of the observer the azimuth of this line is 180°.

At the equinox the latitude of the observer's position could be obtained by letting the upper limb of the sun repeatedly make contact with the horizontal crosswire of the telescope, by plotting the graph as above. The maximum value would then need to be corrected by the semi-diameter of the sun as listed in the Nautical Almanac (correction varies with distance from the sun as the earth's orbit is elliptical) and corrected also for refraction. The value obtained would be the latitude of the observer. In this method care must be taken to make the semi-diameter correction in the right direction, remembering that the sun's image is inverted in the telescope.

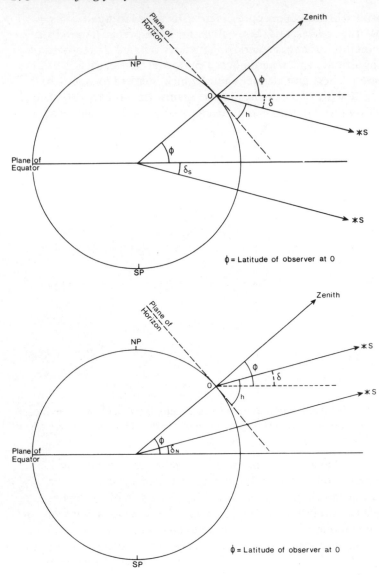

Fig. 109   When the sun is away from the equator (i.e. at all times other than the equinoxes) the latitude can be found by measurement of altitude h when the sun crosses the
observer's meridian. If the sun has South declination, the latitude equals $[90° - (h + \delta)]$
where δ is the declination; if the sun has North declination, the latitude equals
$[90° - (h - \delta)]$.

The method is not very accurate. Firstly, the refraction values are only
approximate, apart from the general disadvantages already listed. Secondly,
the sun will have zero declination at only one specific moment of time,
which will probably not coincide with the L.A.N. of the observer. It is

possible to obtain better results by pairing observations to the sun as described on p. 5, but the non-specialist will find it easier to keep the sun in the telescope if using only the upper limb, when circle readings will change at a steady rate and time will not be lost in searching for the sun's image in the telescope after every change of Face.

The sky is not always clear at the equinoxes. If the method is used on any other date, therefore, the value obtained after correction must be further modified by adding or subtracting the declination of the sun at noon on the day in question, the sign depending on the position of the observer and on the season. Even if the observation is taken at the equinox, the observed latitude must be corrected by the declination of the sun at the observer's L.A.N.

A very rough value for longitude could be obtained by calculating L.M.N. from L.A.N., and then obtaining G.M.N. The G.M.T. of L.A.N. will give a measure of longitude, e.g. if L.M.N. is at 12^h 33^m 20^s G.M.T., it means that Local Mean Noon occurred 33^m 20^s after Mean Noon at Greenwich. 33^m 20^s of time = 08° 20′ 00″ of longitude, so that the observer's longitude would be 08° 20′ 00″ West. However, if time were known to be correct only to the nearest 10 seconds, the longitudinal displacement could be considerable. One difficulty here is the determination of the precise time of L.A.N. from the graph, as the top of the graph is very flat and the precise peak difficult to determine, so that the instant of L.A.N. may not be known more closely than to the nearest minute by watch time, and watch error may itself be known only to the nearest quarter of a minute. A longitude value to the nearest kilometre might therefore represent the limit of accuracy.

Position lines

The position line method of fixation is one which has been in use in navigation for a long time. It makes use only of timed altitudes of heavenly bodies, so that if a theodolite is used only the vertical circle needs to be read. Unlike more accurate methods, no preparatory selection of stars is required; the observer uses any visible and identifiable stars.

A single timed observation of a star of known declination defines a line which can be plotted on a map. A series of observations to different bodies, giving intersecting lines, permits the calculation of position.

In fig. 110 the observer's position O is defined by the circle centred at Q with radius (in arc) of $(90° - h)$, where h is the altitude of the star and Q is the substellar point, i.e. at the time of observation the star was overhead at Q. As the earth rotates, Q does not remain under the star. The substellar point becomes a parallel of latitude, the latitude being equal to the declination of the star. The longitude of Q can be found for the moment of observation, when Q lies in the same meridian plane as the star. In fig. 111, the longitude is $Q\hat{N}G$, where G is Greenwich.

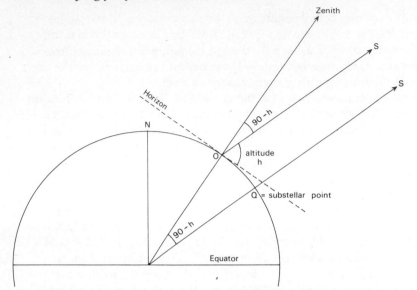

Fig. 110   With the star overhead at Q and the altitude h observed at O, the distance $OQ = 90° - h$. As shown, Q is on the meridian of O, but O could be at any azimuth from Q, at a distance of $90° - h$.

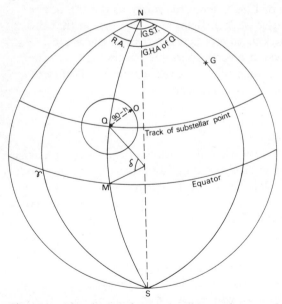

Fig. 111   The observer is at O at a distance from Q of $90° - h$. The star has a Right Ascension shown by the angle R.A. and the arc $♈M$; it has a declination δ shown by the arc MQ and the angle at the centre of the earth.

Greenwich Sidereal Time is the angle G.S.T. between the meridian through G and the meridian of the First Point of Aries $♈$. The Greenwich Hour Angle of Q is the angle G.H.A., i.e. the time since the star crossed the Greenwich Meridian.

The circle on the surface of the earth represents a Position Line through the observer's position. It is distant from Q by a radius $90° - h$ on an azimuth $N\hat{Q}O$, and is at right angles to the radius. The position of Q is defined by R.A., δ and G.S.T.

But

$Q\hat{N}G$ = the Greenwich Hour Angle of the star
= Greenwich Sidereal Time − the Right Ascension of the star.

(Greenwich Sidereal Time is the interval since the lower transit of the 1st Point of Aries across the Greenwich Meridian: the Right Ascension of the star is the angle between the meridian of the 1st Point of Aries and that of the star, i.e. it is the interval between transit of the 1st Point of Aries across the Greenwich meridian and the transit of the star.)

At any point on the position circle, the direction of the circle is at right angles to the radius, i.e. it is perpendicular to the direction from the observer O to the substellar point Q, i.e. it is perpendicular to the direction from O to the star. Wherefore, if the observer chooses a selection of stars in different quadrants, the resulting position lines will give good intersections with each other.

The observer is not totally ignorant of his position. He knows where he is to within a degree or better. His interest therefore lies in the location of the position line in his immediate vicinity. Without a special right-angled eyepiece he will not work stars with a zenith distance of less than 30°, as with most transit theodolites the observer cannot get his eye under the telescope if the altitude exceeds 60°. At Q the zenith distance is zero, the star being overhead. O will therefore be at least 30° of latitude away from Q, so that

the minimum radius of the position circle = (90° − 60°) × (kilometres per degree)

= 30° × 113 kms

= 3390 kms.

Therefore on that section of the chart being used, a segment of the position circle can be regarded as being a straight line.

Calculation of position lines   Initially the observed altitude h is ignored. The latitude ϕ and longitude λ are assumed. The G.M.T. of observation is known, so that Greenwich Sidereal Time can be calculated, and hence the Hour Angle t is calculated.

In the astronomical triangle (see p. 174), the elements $(90 - \phi)$, $(90 - \delta)$ and t are known. The triangle can therefore be solved for azimuth and for altitude corrected for refraction (h).

On the chart, through the assumed position $\lambda\phi$ draw a position line perpendicular to the azimuth line. This is the position line for the altitude h_c. The true observed altitude, corrected for refraction, is h. Therefore a new position line is required, distant from the first by the difference $(h - h_c)$ as in fig. 112. It will be seen from this Figure that h is larger than h_c when h is nearer the star. Wherefore, using the *latitude* scale of the diagram, set out the interval $(h - h_c)$ along the azimuth line in the direction of the star, and draw a new position line parallel to the first, as in fig. 113. It should be clear that if h is smaller than h_c, the intercept $(h - h_c)$ will be negative, and the new position line is moved away from the star instead of towards it.

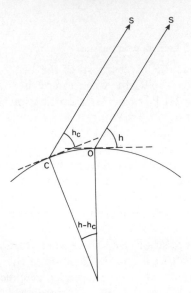

Fig. 112   The observer at O measures the altitude h. For his supposed position C he calculates the altitude h_c. The difference represents the correction to be applied to the position C to bring it to O.

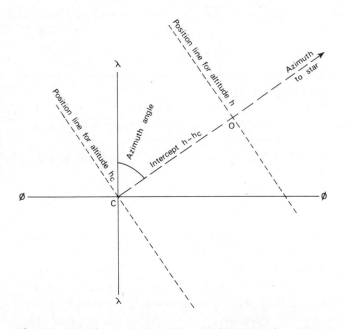

Fig. 113   For the assumed position C, of longitude λ and latitude ϕ, the observer calculates the azimuth to the star and the altitude h_c. As seen in fig. 111, h_c differs from the observed altitude h. The position line, at right angles to the azimuth, is moved from C to O by an amount $h - h_c$, measured on the scale of the chart or map being used.

Position lines for all the observed stars are plotted on the same chart. Each line should have its azimuth direction shown as in fig. 114. A circle is drawn as nearly as possible to touch all the position lines as shown. The centre of the circle can be assumed to be the observer's position.

Results are improved if four stars are used in place of two or three, and if the intersection angles are good. Each star observed is normally taken on Face Left and Face Right, the adopted time values and altitude values

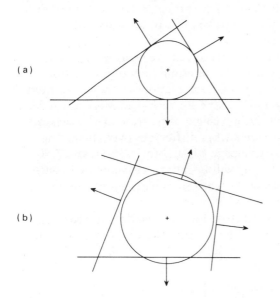

(a)

(b)

Fig. 114   Location of the observer in relation to Position Lines.
(a) Three Position Lines
(b) Four Position Lines
    Azimuth directions are shown in both cases. A circle is drawn tangential to all lines, or as nearly so as possible, and its centre is assumed to be the observer's location.

being the means of the readings on the two Faces. There is always the possibility of an error in recognition on one Face, with a reading to the wrong star: if three or more timed intersections are made, within a space of four minutes, on one Face only, this acts as a quick check on reliability and makes identification error on the other Face immediately recognisable. If possible, the four stars used should lie within an altitude range not exceeding 10°; this reduces errors due to refraction. Finally, approximate values for azimuths can be obtained by setting the horizontal circle to read zero while sighting on Polaris, and taking one horizontal circle reading to each star in the course of observation. This reading should agree with the calculated azimuth value to within $1\frac{1}{2}°$, and checks the computation. The circle readings also provide an immediate check on intersection angles between position lines. They can simplify intersection of a star after change of

Face, by alignment of the telescope on a circle reading differing by 180°
from the preceding reading, with identification of the star following from
vertical movement only.

Position lines from the sun   The non-specialist may hesitate to attempt
observational work with the theodolite in the dark, and may be uncertain
of identifying stars correctly. The method may still be used, however, with
the substitution of sun sights for star sights, although the accuracy of the
results is very much reduced. For navigational purposes, a series of star
sights in quick succession is necessary for the determination of approximate position of the moving ship or aircraft, but the geographer standing
on the ground can spend all day over the determination of his position
if he wishes. He can therefore take paired observations to the sun at intervals
throughout the day, drawing a position line for each pair. Morning, midday
and afternoon observations will give a triangle of position lines which can
be used as above, although there are always difficulties over attempting
azimuth calculations at mid-day (see page 13) and four sets at, say, 7.30,
10.30, 13.30 and 16.30 hours L.A.T. will be more satisfactory. Of course,
altitudes will not be as well balanced as with star observations, nor are
the azimuths directed into each quadrant.

Calculation is easier than with stars, as sidereal time does not come into
the computation. The declination of the sun must be calculated for each
position line, as it changes relatively rapidly.

The Greenwich Hour Angle of the sun = G.A.T. of the observation

$$= \text{G.M.T.} \pm \text{Equation of Time}$$

$$= \text{L.M.T.} \pm \text{Equation of Time}$$
$$\pm \text{Longitude in Time}$$

The computation of solar position line observations follows previous
procedure. A time check is made, using approximate latitude ϕ. This gives
the calculated value of Local Apparent Time. A comparison of this with
G.M.T., obtained from a watch checked with wireless signals, adjusted for
the Equation of Time to give G.A.T., provides approximate longitude λ. e.g.

observation Sept. 26 at Watch Time	$10^{\text{h}}\ 36^{\text{m}}\ 00^{\text{s}}$	$(= \text{G.M.T.} + 1^{\text{h}})$
calculated L.A.T.	$09^{\text{h}}\ 32^{\text{m}}\ 00^{\text{s}}$	
Equation of Time	$-\quad 08^{\text{m}}\ 20^{\text{s}}$	
L.M.T.	$= \text{L.A.T.} + \text{Equation of Time}$	
	$= 09^{\text{h}}\ 32^{\text{m}}\ 00^{\text{s}} - 08^{\text{m}}\ 20^{\text{s}}$	
	$= 09^{\text{h}}\ 23^{\text{m}}\ 40^{\text{s}}$	
Watch error on G.M.T.	$=\quad\ 07^{\text{m}}\ 30^{\text{s}}$	(by comparison with radio signals)

$$\text{G.M.T.} \qquad = 09^h\ 36^m\ 00^s - 07^m\ 30^s$$
$$= 09^h\ 28^m\ 30^s$$
$$\text{G.M.T.} - \text{L.M.T.} = \qquad 04^m\ 50^s$$
$$01^h\ 00^m\ 00^s \quad = \ 15°\ \text{longitude}$$
$$04^m\ 50^s \quad = \ 01°\ 12'\ 30''\ \text{longitude}$$

As L.M.T. is less than G.M.T., longitude is West
$$= 01°\ 12'\ 30''\ \text{West.}$$

This figure can be taken as the approximate longitude, and used with an approximate value of latitude. The azimuth is calculated in the normal way (see p. 8). The assumed parallel of latitude is plotted on the graph paper and longitudes are marked off to scale.

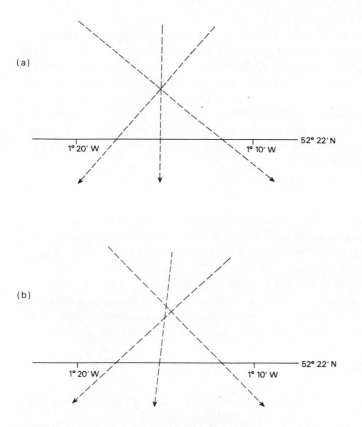

Fig. 115 Azimuth lines from sun observations.
(a) Even if the assumed latitude were incorrect, the lines from three sets of observations would meet at a point if the azimuths were correct.
(b) The error in latitude will throw all azimuths slightly into error so that the lines will form a triangle of error. The desired point will be outside the triangle and will be the centre of a circle tangential to all three position lines.

If the assumed *latitude is exactly correct*, then in the astronomical tri-angle $(90 - \phi)$ is correct and $(90 - b)$ is correct (the observed value b equals b_c). The variable quantities are

t	(hour angle of the sun)
$(90 - \delta)$	(polar distance, dependent on t)
azimuth angle	(dependent on t)

If the assumed *latitude is incorrect,* then $(90 - \phi)$ is slightly in error, but $(90 - b)$ remains correct. If the watch is correct, t will be slightly in error (as latitude is used in its calculation) and $(90 - \delta)$ will also be in error. This will give slightly different longitude values on the assumed parallel of latitude. If the azimuths were correct the graph would appear as in fig. 115(a), the three azimuth lines meeting at a point on a latitude slightly different from that assumed, and both latitude and longitude could be read from the graph. However, the error in assumed latitude will also throw all azimuths slightly into error, so that they will not meet at a point but will form a triangle of error, as in fig. 115(b). The desired point will be *outside* the triangle in the sector to the left or to the right of all three rays (compare plane table resection on p. 92). Only one of the two possible sectors will be bounded by all three rays. To find the precise location, bisect the central segment at right angles and find on it the centre of a circle which will be tangential to all three rays.

Plotting of position lines   Note that whether we are plotting stellar or solar position lines it is assumed that it is possible to draw azimuths and position lines as straight lines set off from North with a protractor. This implies that the chart being used is on a Mercator projection. On this projection the scale at any point is the same in all directions, i.e., 1000 m East–West will measure the same as 1000 m North–South. Geographical positions, however, are given in degrees, minutes and seconds of arc, and away from the equator a degree of latitude is larger than a degree of

longitude. A degree of latitude on the globe is $\dfrac{2\pi R}{360}$ where R is the radius of

the earth. A degree of longitude is $\dfrac{2\pi R \cdot \cos \phi}{360}$ where ϕ is the latitude of the

parallel being considered.

In plotting the chart on which the position lines are to be drawn, the scale used in setting out the graduations of longitude depends upon the results obtained from observations and the accuracy of these observations. Having selected the longitude scale, the latitude scale must be calculated. At latitude

ϕ the length of a parallel on the globe is $2\pi R . \cos \phi$ and one minute of longi-

tude will be $\dfrac{2\pi R . \cos \phi}{360 . 60}$. On meridians the length of a degree is everywhere

constant, so that one degree of latitude is $\dfrac{2\pi R}{360}$ and one minute of latitude

will be $\dfrac{2\pi R}{360 . 60}$.

The ratio of longitude scale to latitude scale is therefore

$$\frac{2\pi R . \cos \phi}{360 . 60} : \frac{2\pi R}{360 . 60}$$

$$= \quad \cos \phi : 1$$

$$= \quad 1 : \frac{1}{\cos \phi}$$

$$= \quad 1 : \sec \phi$$

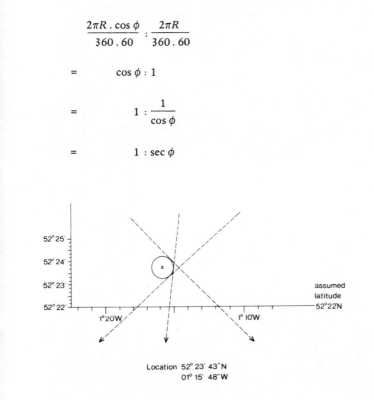

Fig. 116   Plotting of solar azimuth lines. The latitude scale is calculated as described in the text.

Because the three azimuths differ by less than 180°, the circle whose centre is taken as the observer's position is outside the triangle of error, which arises from shift in azimuth following use of an incorrect latitude value.

This method of plotting can be used with star sights in place of the intercept method. If four stars in different quadrants are observed, the circle lies inside the quadrilateral formed by the azimuth lines.

In the example used, $\phi = 52°\ 22'$ N.; whence $\sec \phi = 1.638$. If the graph is drawn with a scale of 10 millimetres for one minute of longitude, one minute of latitude will be 16.38 millimetres. The vertical scale can therefore be plotted and the latitude of the centre of the circle can be measured as in fig. 116.

This method of plotting can be used equally well with star observations, in place of the intercept method described earlier.

Appendix 1
The geometry of resection

Trigonometrical resection (Collins Point)

It is sometimes easier to determine the co-ordinates of an unknown point by trigonometrical resection, measuring angles with a theodolite, than it is to carry out a theodolite traverse. If previously fixed points are marked with visible poles, tripods or other signals, one observer working by himself can complete the fixation.

The usual form of resection is similar to that achieved graphically with the plane table (see p. 90). Observations are taken to three stations whose positions are already known. In fig. 117, the positions of stations A, B and C are known. The observer sets the theodolite over the required station P, and measures angles α and β in the usual way, i.e. with FL and FR observations, using one station as a Reference Object for starting and finishing each round, and calculating the angles as the mean values obtained from several sets of observations

In fig. 117,

$A\hat{P}C$ is observed = α
$C\hat{P}B$ is observed = β

Calculation

An angle $(180° - \alpha)$ can be constructed at B to meet CP produced at Q. Then

$Q\hat{B}A = 180 - \alpha = A\hat{P}Q$
$Q\hat{B}A$ and $Q\hat{P}A$ stand opposite the same side AQ and are equal.
∴ A, P, B and Q are all on a circle of which AQ is a chord.

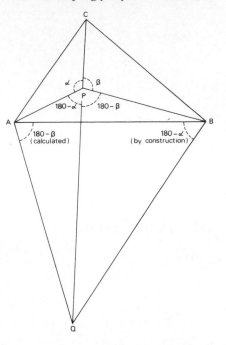

Fig. 117   Trigonometrical Resection.
    The observer is at an unknown point P. He can see three known points A, B and C, and is inside the triangle ABC. He observes the angles α and β and calculates his position (see the text).

Then

> BQ is also a chord, and
> $B\hat{A}Q = B\hat{P}Q = 180 - \beta$

Wherefore:

1. The azimuth AB and the length AB are known.
2. The triangle ABQ can be calculated, as side AB and the two angles $B\hat{A}Q$ and $Q\hat{B}A$ are known, using the sine rule. The co-ordinates of Q are therefore found, as the length AQ and the azimuth AQ are calculated, and the co-ordinates of A are known. The calculation is checked by working from B with the length BQ and the azimuth BQ.
3. As the co-ordinates of C and Q are now known, the azimuth CQ can be calculated.
4. The triangle APQ can now be solved, as

> AQ is known from step 2
> $Q\hat{P}A = 180 - \alpha$
> $A\hat{Q}P = (\text{azimuth } PQ) - (\text{azimuth } AQ)$

The co-ordinates of P can therefore be calculated, as the azimuth AP and the distance AP can both be computed. A check is made by calculating also through the triangle BPQ.

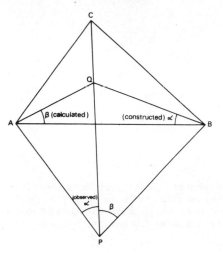

Fig. 118   Trigonometrical Resection.
   As for fig. 117, but with the observer outside the triangle formed by the three known points A, B and C. He measures the angles α and β.

Figure 118 shows a similar fixation but with P outside the triangle ABC. In this case the angle constructed at B = α and not (180 − α). Likewise the calculated value of QAB = β and not (180 − β). The computation remains virtually the same.

Note: If in this second case P is concentric with A, B and C, the calculation is impossible, as Q coincides with B.

This form of resection is known as the Collins Point method. There are many other ways of completing resections: individual surveyors tend to have personal preferences.

Plane table (Bessel's method)

Procedure:
   1. Set up the plane table at P, from which A, B and C are visible.
   2. Set the alidade from b to a and sight to A. Clamp the table.
   3. Draw a ray C_b from b to C. Unclamp the table.
   4. Set the alidade from a to b and sight to B. Clamp the table.
   5. Draw a ray C_a from a to C. Unclamp the table.
   6. Let the rays aC_a and bC_b intersect at d. Set the alidade from d to c and sight to C. Clamp the table.

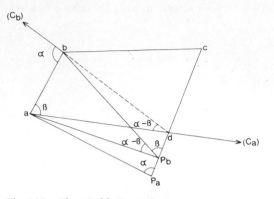

Fig. 119   Plane Table Resection.
  Diagram for Bessel's method, described in the text. C_a and C_b are the two rays drawn by the observer at the unknown position. They intersect at d on the sheet. The plane table is oriented on the line dc and is clamped. Back rays are drawn from a and b to intersect at p, which is the observer's position. The text shows that only one position is possible for p.

7. The table is now oriented. Pivot the alidade round a and sight on A. Draw a ray back from a to meet cd at p_a.
8. Pivot the alidade round b and sight on B. Draw a ray back from b to meet cd at p_b.

Proof

Let us assume that the two final rays ap_a and bp_b intersect cd at p_a and p_b respectively.

By construction,

$$a\widehat{b}C_b \text{ (drawn on the board)} = A\widehat{P}C \text{ (existing on the ground)} = \alpha$$

$$a\widehat{p}_a c \text{ (existing on the board)} = A\widehat{P}C = \alpha$$

$$b\widehat{a}C_a \text{ (drawn on the board)} = B\widehat{P}C \text{ (existing on the ground)} = \beta$$

$$b\widehat{p}_b c \text{ (existing on the board)} = B\widehat{P}C = \beta$$

In the quadrilateral $abdp_b$, the two angles $b\widehat{a}d$ and $b\widehat{p}_b d = \beta$. As they stand on the same side bd, bd is a chord of a circle in which $abdp_b$ is a circumscribed quadrilateral.

By simple geometry, $a\widehat{b}C_b = b\widehat{a}d + a\widehat{d}b$

$$a\widehat{d}b = a\widehat{b}C_b - b\widehat{a}d = \alpha - \beta$$

As $abdp_b$ are concyclic, $a\widehat{p}_b b = a\widehat{d}b = \alpha - \beta$

Then $a\widehat{p}_b c = a\widehat{p}_b b + b\widehat{p}_b c = (\alpha - \beta) + \beta = \alpha$

But $a\widehat{p}_a c = \alpha$ by construction

$$a\widehat{p}_b c = a\widehat{p}_a c + p_b \widehat{a} p_a$$

∴ $p_b a p_a$ = zero

and as p_a, p_b, d and c are on a straight line p_a and p_b must coincide.

∴ the single point p subtends with a, b and c the same angles that P subtends with A, B and C, and so p correctly determines the position P.

Note: If A, B, C and P are concyclic, resection is impossible. a, b, c and p must be concyclic, but a, b, d and p are also concyclic and the line pdc is straight. Then c and d must coincide, and the line cd has no length: it is therefore impossible to orient the board on the line from d to c. The two rays $C_b b$ and aC_a will intersect at c.

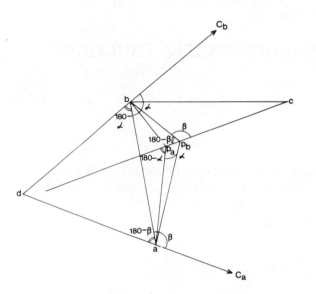

Fig. 120   Plane Table Resection.
    As for fig. 119, but with p inside the triangle abc.

In the example given the point P lay outside the triangle formed by the three known points A, B and C. A similar proof holds good if P is inside the triangle ABC.

Appendix 2
Basic trigonometric functions

In any right-angled triangle, the ratios between the side lengths can be defined. In the triangle XOP, with the angle $X\hat{O}P = \theta$, the ratios are:

$$\frac{XP}{OP} = \sin \theta \quad \text{and} \quad \frac{1}{\sin \theta} = \operatorname{cosec} \theta$$

$$\frac{XO}{OP} = \cos \theta \quad \text{and} \quad \frac{1}{\cos \theta} = \sec \theta$$

$$\frac{XP}{XO} = \tan \theta \quad \text{and} \quad \frac{1}{\tan \theta} = \cot \theta$$

In fig. 123 there are a number of triangles $X_1 P_1 O$, $X_2 P_2 O$, etc., which are of identical dimensions albeit differently oriented. The bearing OP_1 (or OP_2 etc.) is the angle $N\hat{O}P_1$ (or $N\hat{O}P_2$ etc.).

If one considers the two triangles $X_1 P_1 O$ and $X_2 P_2 O$

$$OP_1 = OP_2$$
$$X_1 \hat{O} P_1 = X_2 \hat{P}_2 O \quad (= 30° \text{ in this case})$$
$$X_1 \hat{P}_1 = X_2 O$$
$$X_1 \hat{P}_1 O = X_2 \hat{O} P_2 \quad (= 60° \text{ in this case})$$
$$\sin X_1 \hat{O} P_1 = \text{sine of bearing } OP_1 = \frac{X_1 P_1}{OP_1}$$
$$= \sin 30°$$

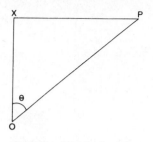

Fig. 121   The basic right-angled triangle used in simple plane trigonometry.

and must equal

$$\sin X_2 \hat{P}_2 O = \frac{X_2 O}{OP_2}$$

$$= \cos X_2 \hat{O} P_2$$

$$= \cos 60°$$

Similarly one can show that

$$\sin 30° \quad = + \cos 60°$$

$$= - \cos 120°$$

$$= + \sin 150°$$

$$= - \sin 210°$$

$$= - \cos 240°$$

$$= + \cos 300°$$

$$= - \sin 330°$$

The + and − signs can be understood by comparison of figs. 122 and 123, giving appropriate sign to $X_n P_n$ and $X_n O$ in each triangle. In the case of $X_3 P_3$, it can be seen that $X_3 P_3 = OP_3$, i.e. $\sin 90° = 1$ and $\cos 90° =$ zero.

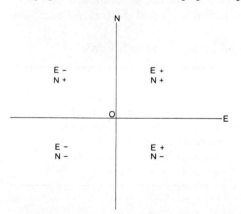

Fig. 122   North–South and East–West axes are shown through the observer's position at O. All points East of O have positive Eastings; points West of O have negative Eastings. Similarly points North of O have positive Northings; those South of O have negative Northings.

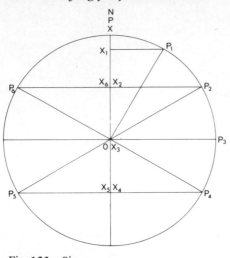

Fig. 123   Sines.

$$\text{Sin } X\hat{O}P = \frac{XP}{OP}$$

All trigonometrical functions can be expressed in terms of some angle between 0° and 45°, and this is what is done in reference tables. It is important to consider the sign when calculating co-ordinates. If uncertain, consider any line, e.g. in a theodolite traverse, in terms of figs. 122 and 123. For example, the line OP_4 has a bearing of 120° so that P_4 will have

    +ve Eastings   −ve Northings

relative to 0; P_5 has a bearing of 240° from 0 and will have

    −ve Eastings   −ve Northings

relative to 0. As 240° is more West than South, the difference in Eastings must be greater than the difference in Northings and the traverse computation should show this.

Basic trigonometrical functions can be tabulated as follows:

Bearing or Angle	sin	cos	tan
0°–90°	Between 0 and +1	Between +1 and 0	Between 0 and +
90°–180°	Between +1 and 0	Between 0 and −1	Between − and 0
180°–270°	Between 0 and −1	·Between −1 and 0	Between 0 and +
270°–360°	Between −1 and 0	Between 0 and +1	Between − and 0

Tables

Tables of trigonometrical functions therefore list values of sine, cosine, tangent and cotangent only for angles between 0° and 45°. Some, such as Shortrede, also number these to show functions from 45° to 360°. When using tables such as Chambers, numbered only from 0° to 45°, one needs to convert the desired function to a corresponding statement for an angle in this range.

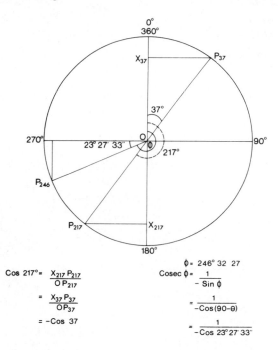

$$\cos 217° = \frac{X_{217}\,P_{217}}{O\,P_{217}}$$

$$= \frac{X_{37}\,P_{37}}{OP_{37}}$$

$$= -\cos 37$$

$$\phi = 246°\;32\;27$$

$$\operatorname{Cosec}\phi = \frac{1}{-\sin\phi}$$

$$= \frac{1}{-\cos(90-\theta)}$$

$$= \frac{1}{-\cos 23°\,27'\,33''}$$

Fig. 124 To keep tables of trigonometrical functions as short as possible, only the range 0°–45° is included. Any function of an angle greater than 45° is converted to a function of an angle in the table.

For example, if it is desired to find cos 217° as shown in Figure 124, one seeks the angle between the bearing given (217°) and the nearest N–S or E–W axis, in this case the N–S axis at 180°. Subtracting 180° from 217°, the angle is 37°, and cos 217° = cos 37°.

As a second example, if one needs cosec 246° 32' 27", the nearest axis is at 270°, the angular difference being 23° 27' 33". As Figure 136 shows,

$$\operatorname{cosec} 246°\,32'\,27'' = -\frac{1}{\cos 23°\,27'\,33''}$$

Reciprocals are best calculated logarithmically. In the latter case

$$\cos 23°\,27'\,33'' = \bar{1}.962\,5322$$

Subtract this from 0 (which is the logarithm of 1), to give

$$0.037\,4678$$

and the antilogarithm = 1.090 1036

In this case the answer needs a −ve sign.

Following the convention that 10 is added to a negative characteristic of a logarithm

$$\log \cos 23° \, 27' \, 33'' = 9.962 \, 5322$$

Subtract from 10   = 0.037 4678
Antilogarithm   = 1.090 1036

$$= \operatorname{cosec} 246° \, 32' \, 27'' \text{ as before.}$$

Topographical
Survey Tables

TABLE 1   Sun's parallax in altitude

Altitude	Parallax	Altitude	Parallax
10°	9″	35°	7″
15°	8″	40°	7″
20°	8″	45°	6″
25°	8″	50°	6″
30°	8″	55°	5″

TABLE 2   Astronomical refraction. For 30 ins/1016 mb pressure and 50°F./9°C.

Altitude	Refraction	Altitude	Refraction	Altitude	Refraction
10° 00′	5′ 20″	14° 00′	3′ 50″	30° 00′	1′ 41″
10′	5′ 15″	10′	3′ 47″	31° 00′	1′ 37″
20′	5′ 10″	20′	3′ 44″	32° 00′	1′ 33″
30′	5′ 05″	30′	3′ 42″	33° 00′	1′ 29″
40′	5′ 00″	40′	3′ 39″	34° 00′	1′ 26″
50′	4′ 56″	50′	3′ 37″	35° 00′	1′ 23″
11° 00′	4′ 51″	15° 00′	3′ 33″	36° 00′	1′ 20″
10′	4′ 47″	30′	3′ 27″	37° 00′	1′ 17″
20′	4′ 43″	16° 00′	3′ 21″	38° 00′	1′ 14″
30′	4′ 39″	30′	3′ 14″	39° 00′	1′ 12″
40′	4′ 35″	17° 00′	3′ 08″	40° 00′	1′ 09″
50′	4′ 31″	30′	3′ 03″	41° 00′	1′ 07″
12° 00′	4′ 28″	18° 00′	2′ 58″	42° 00′	1′ 05″
10′	4′ 24″	19° 00′	2′ 48″	43° 00′	1′ 02″
20′	4′ 21″	20° 00′	2′ 39″	44° 00′	1′ 00″
30′	4′ 17″	21° 00′	2′ 31″	45° 00′	0′ 58″
40′	4′ 14″	22° 00′	2′ 23″	46° 00′	0′ 56″
50′	4′ 11″	23° 00′	2′ 16″	47° 00′	0′ 54″
13° 00′	4′ 07″	24° 00′	2′ 10″	48° 00′	0′ 52″
10′	4′ 04″	25° 00′	2′ 04″	49° 00′	0′ 51″
20′	4′ 01″	26° 00′	1′ 59″	50° 00′	0′ 49″
30′	3′ 58″	27° 00′	1′ 54″	51° 00′	0′ 47″
40′	3′ 55″	28° 00′	1′ 49″	52° 00′	0′ 45″
50′	3′ 53″	29° 00′	1′ 45″	53° 00′	0′ 44″

TABLE 3   Corrections for slope for measured distances of 100 units

Degrees	Minutes					
	00	10	20	30	40	50
00	00.00	00.00	00.00	00.00	00.01	00.01
01	00.02	00.02	00.03	00.03	00.04	00.05
02	00.06	00.07	00.08	00.10	00.11	00.12
03	00.14	00.15	00.17	00.19	00.20	00.22
04	00.24	00.26	00.29	00.31	00.33	00.36
05	00.38	00.41	00.43	00.46	00.49	00.52
06	00.55	00.58	00.61	00.64	00.68	00.71
07	00.75	00.78	00.82	00.86	00.89	00.93
08	00.97	01.01	01.06	01.10	01.14	01.19
09	01.23	01.28	01.32	01.37	01.42	01.47
10	01.52	01.57	01.62	01.67	01.73	01.78
11	01.84	01.89	01.95	02.01	02.07	02.13
12	02.19	02.25	02.31	02.37	02.43	02.50
13	02.56	02.63	02.70	02.76	02.83	02.90
14	02.97	03.04	03.11	03.19	03.26	03.33
15	03.41	03.48	03.56	03.64	03.72	03.79
16	03.87	03.95	04.04	04.12	04.20	04.28
17	04.37	04.45	04.54	04.63	04.72	04.80
18	04.89	04.98	05.08	05.17	05.26	05.35
19	05.45	05.54	05.64	05.74	05.83	05.93
20	06.03	06.13	06.23	06.33	06.44	06.54
21	06.64	06.75	06.85	06.96	07.07	07.17
22	07.28	07.39	07.50	07.61	07.72	07.84
23	07.95	08.06	08.18	08.29	08.41	08.53
24	08.65	08.76	08.88	09.00	09.12	09.25
25	09.37	09.49	09.62	09.74	09.87	09.99
26	10.12	10.25	10.38	10.51	10.64	10.77
27	10.90	11.03	11.16	11.30	11.43	11.57
28	11.71	11.84	11.98	12.12	12.26	12.40
29	12.54	12.68	12.82	12.96	13.11	13.25
30	13.40	13.54	13.69	13.84	13.99	14.13
31	14.28	14.43	14.58	14.74	14.89	15.04
32	15.20	15.35	15.50	15.66	15.82	15.97
33	16.13	16.29	16.45	16.61	16.77	16.93
34	17.10	17.26	17.42	17.59	17.75	17.92
35	18.08	18.25	18.42	18.59	18.76	18.93
36	19.10	19.27	19.44	19.61	19.79	19.96
37	20.14	20.31	20.49	20.66	20.84	21.02
38	21.20	21.38	21.56	21.74	21.92	22.10
39	22.29	22.47	22.65	22.84	23.02	23.21
40	23.40	23.58	23.77	23.96	24.15	24.34

Example of use of the table:
   The line *AB* is measured as follows:
300.00 m with slope of 03° 10′.   Correction = 3   x 0.15 = 0.45
150.00 m with slope of 04° 10′.   = 1.5   x 0.26 = 0.39
275.55 m with slope of 02° 50′.   = 2.75 x 0.12 = 0.33
725.55 m   Correction =   1.17

True horizontal distance = 724.38 m

TABLE 4A   True horizontal distances from stadia intercepts

Angle ° '	Apparent Slant Range								
	100	200	300	400	500	600	700	800	900
00 00	100.00	200.00	300.00	400.00	500.00	600.00	700.00	800.00	900.00
10	100.00	200.00	300.00	400.00	500.00	599.99	699.99	799.99	899.99
20	100.00	199.99	299.99	399.99	499.98	599.98	699.98	799.97	899.97
30	99.99	199.98	299.98	399.97	499.96	599.95	699.95	799.94	899.93
40	99.99	199.97	299.96	399.95	499.93	599.92	699.91	799.89	899.88
50	99.98	199.96	299.94	399.92	499.89	599.87	699.85	799.83	899.81
01 00	99.97	199.94	299.91	399.88	499.85	599.82	699.79	799.76	899.73
10	99.96	199.92	299.88	399.83	499.79	599.75	699.71	799.67	899.63
20	99.95	199.89	299.84	399.78	499.73	599.68	699.62	799.57	899.51
30	99.93	199.86	299.79	399.73	499.66	599.59	699.52	799.45	899.38
40	99.92	199.83	299.75	399.66	499.58	599.49	699.41	799.32	899.24
50	99.90	199.80	299.69	399.59	499.49	599.39	699.28	799.18	899.08
02 00	99.88	199.76	299.63	399.51	499.39	599.27	699.15	799.03	898.90
10	99.86	199.71	299.57	399.43	499.29	599.14	699.00	798.86	898.71
20	99.83	199.67	299.50	399.34	499.17	599.01	698.84	798.67	898.51
30	99.81	199.62	299.43	399.24	499.05	598.86	698.67	798.48	898.29
40	99.78	199.57	299.35	399.13	498.92	598.70	698.48	798.27	898.05
50	99.76	199.51	299.27	399.02	498.78	598.53	698.29	798.05	897.80
03 00	99.73	199.45	299.18	398.90	498.63	598.36	698.08	797.81	897.53
10	99.69	199.39	299.08	398.78	498.47	598.17	697.86	797.56	897.25
20	99.66	199.32	298.99	398.65	498.31	597.97	697.63	797.30	896.96
30	99.63	199.25	298.88	398.51	498.14	597.76	697.39	797.02	896.65
40	99.59	199.18	298.77	398.36	497.96	597.55	697.14	796.73	896.32
50	99.55	199.11	298.66	398.21	497.77	597.32	696.87	796.42	895.98
04 00	99.51	199.03	298.54	398.05	497.57	597.08	696.59	796.11	895.62
10	99.47	198.94	298.42	397.89	497.36	596.83	696.30	795.78	895.25
20	99.43	198.86	298.29	397.72	497.15	596.57	696.00	795.43	894.86
30	99.38	198.77	298.15	397.54	496.92	596.31	695.69	795.08	894.46
40	99.34	198.68	298.01	397.35	496.69	596.03	695.37	794.70	894.04
50	99.29	198.58	297.87	397.16	496.45	595.74	695.03	794.32	893.61
05 00	99.24	198.48	297.72	396.96	496.20	595.44	694.68	793.92	893.16
10	99.19	198.38	297.57	396.76	495.95	595.13	694.32	793.51	892.70
20	99.14	198.27	297.41	396.54	495.68	594.82	693.95	793.09	892.22
30	99.08	198.16	297.24	396.33	495.41	594.49	693.57	792.65	891.73
40	99.03	198.05	297.07	396.10	495.13	594.15	693.18	792.20	891.23
50	98.97	197.93	296.90	395.87	494.84	593.80	692.77	791.74	890.70
06 00	98.91	197.81	296.72	395.63	494.54	593.44	692.35	791.26	890.17
10	98.85	197.69	296.54	395.38	494.23	593.08	691.92	790.77	889.61
20	98.78	197.57	296.35	395.13	493.92	592.70	691.48	790.26	889.05
30	98.72	197.44	296.16	394.87	493.59	592.31	691.03	789.75	888.47
40	98.65	197.30	295.96	394.61	493.26	591.91	690.57	789.22	887.87
50	98.58	197.17	295.75	394.34	492.92	591.51	690.09	788.67	887.26
07 00	98.51	197.03	295.54	394.06	492.57	591.09	689.60	788.12	886.63
10	98.44	196.89	295.33	393.77	492.22	590.66	689.11	787.55	885.99
20	98.37	196.74	295.11	393.48	491.85	590.22	688.60	786.97	885.34
30	98.30	196.59	294.89	393.19	491.48	589.78	688.07	786.37	884.67
40	98.22	196.44	294.66	392.88	491.10	589.32	687.54	785.76	883.98
50	98.14	196.28	294.43	392.57	490.71	588.85	687.00	785.14	883.28
08 00	98.06	196.13	294.19	392.25	490.32	588.38	686.44	784.50	882.57
10	97.98	195.96	293.95	391.93	489.91	587.89	685.87	783.86	881.84
20	97.90	195.80	293.70	391.60	489.50	587.40	685.30	783.20	881.10
30	97.82	195.63	293.45	391.26	489.08	586.89	684.71	782.52	880.34
40	97.73	195.46	293.19	390.92	488.65	586.38	684.11	781.83	879.56
50	97.64	195.28	292.93	390.57	488.21	585.85	683.49	781.14	878.78
09 00	97.55	195.11	292.66	390.21	487.76	585.32	682.87	780.42	877.98
10	97.46	194.92	292.39	389.85	487.31	584.77	682.23	779.70	877.16
20	97.37	194.74	292.11	389.48	486.85	584.22	681.59	778.96	876.33
30	97.28	194.55	291.83	389.10	486.38	583.66	680.93	778.21	875.48
40	97.18	194.36	291.54	388.72	485.90	583.08	680.26	777.44	874.62
50	97.08	194.17	291.25	388.33	485.42	582.50	679.58	776.67	873.75

TABLE 4A—*continued*

Angle ° '	Apparent Slant Range								
	100	200	300	400	500	600	700	800	900
10 00	96.98	193.97	290.95	387.94	484.92	581.91	678.89	775.88	872.86
10	96.88	193.77	290.65	387.54	484.42	581.30	678.19	775.07	871.96
20	96.78	193.56	290.35	387.13	483.91	580.69	677.48	774.26	871.04
30	96.68	193.36	290.04	386.72	483.40	580.07	676.75	773.43	870.11
40	96.57	193.15	289.72	386.30	482.87	579.44	676.02	772.59	869.17
50	96.47	192.93	289.40	385.87	482.34	578.80	675.27	771.74	868.21
11 00	96.36	192.72	289.08	385.44	481.80	578.16	674.51	770.87	867.23
10	96.25	192.50	288.75	385.00	481.25	577.50	673.75	770.00	866.24
20	96.14	192.28	288.41	384.55	480.69	576.83	672.97	769.10	865.24
30	96.03	192.05	288.08	384.10	480.13	576.15	672.18	768.20	864.23
40	95.91	191.82	287.73	383.64	479.55	575.46	671.38	767.29	863.20
50	95.79	191.59	287.38	383.18	478.97	574.77	670.56	766.36	862.15
12 00	95.68	191.35	287.03	382.71	478.39	574.06	669.74	765.42	861.10
10	95.56	191.12	286.67	382.23	477.79	573.35	668.91	764.47	860.02
20	95.44	190.88	286.31	381.75	477.19	572.63	668.06	763.50	858.94
30	95.32	190.63	285.95	381.26	476.58	571.89	667.21	762.52	857.84
40	95.19	190.38	285.57	380.77	475.96	571.15	666.34	761.53	856.73
50	95.07	190.13	285.20	380.27	475.33	570.40	665.47	760.53	855.60
13 00	94.94	189.88	284.82	379.76	474.70	569.64	664.58	759.52	854.46
10	94.81	189.62	284.43	379.25	474.06	568.87	663.68	758.49	853.30
20	94.68	189.36	284.04	378.73	473.41	568.09	662.77	757.45	852.13
30	94.55	189.10	283.65	378.20	472.75	567.30	661.85	756.40	850.95
40	94.42	188.84	283.25	377.67	472.09	566.61	660.92	755.34	849.76
50	94.28	188.57	282.85	377.13	471.42	565.70	659.98	754.27	848.55
14 00	94.15	188.29	282.44	376.59	470.74	564.88	659.03	753.18	847.33
10	94.01	188.02	282.03	376.04	470.05	564.06	658.07	752.08	846.09
20	93.87	187.74	281.61	375.49	469.36	563.23	657.10	750.97	844.84
30	93.73	187.46	281.19	374.92	468.65	562.39	656.12	749.85	843.58
40	93.59	187.18	280.77	374.36	467.95	561.54	655.12	748.71	842.30
50	93.45	186.89	280.34	373.78	467.23	560.68	654.12	747.57	841.01
15 00	93.30	186.60	279.90	373.21	466.51	559.81	653.11	746.41	839.71
10	93.16	186.31	279.47	372.62	465.78	558.93	652.09	745.24	838.40
20	93.01	186.01	279.02	372.03	465.04	558.04	651.05	744.06	837.07
30	92.86	185.72	278.57	371.43	464.29	557.15	650.01	742.87	835.73
40	92.71	185.42	278.12	370.83	463.54	556.25	648.95	741.66	834.37
50	92.56	185.11	277.67	370.22	462.78	555.33	647.89	740.45	833.00
16 00	92.40	184.80	277.21	369.61	462.01	554.41	646.82	739.22	831.62
10	92.25	184.50	276.74	368.99	461.24	553.49	645.73	737.98	830.23
20	92.09	184.18	276.27	368.36	460.46	552.55	644.64	736.73	828.82
30	91.93	183.87	275.80	367.73	459.67	551.60	643.53	735.47	827.40
40	91.77	183.55	275.32	367.10	458.87	550.65	642.42	734.20	825.97
50	91.61	183.23	274.84	366.46	458.07	549.68	641.30	732.91	824.52
17 00	91.45	182.90	274.36	365.81	457.26	548.71	640.16	731.61	823.07
10	91.29	182.58	273.87	365.15	456.44	547.73	639.02	730.31	821.60
20	91.12	182.25	273.37	364.49	455.62	546.74	637.87	728.99	820.11
30	90.96	181.92	272.87	363.83	454.79	545.75	636.70	727.66	818.62
40	90.79	181.58	272.37	363.16	453.95	544.74	635.53	726.32	817.11
50	90.62	181.24	271.86	362.48	453.11	543.73	634.35	724.97	815.59
18 00	90.45	180.90	271.35	361.80	452.25	542.71	633.16	723.61	814.06
10	90.28	180.56	270.84	361.12	451.40	541.68	631.95	722.23	812.51
20	90.11	180.21	270.32	360.42	450.53	540.64	630.74	720.85	810.96
30	89.93	179.86	269.80	359.73	449.66	539.59	629.52	719.45	809.39
40	89.76	179.51	269.27	359.02	448.78	538.54	628.29	718.05	807.80
50	89.58	179.16	268.74	358.32	447.89	537.47	627.05	716.63	806.21
19 00	89.40	178.80	268.20	357.60	447.00	536.40	625.80	715.20	804.60
10	89.22	178.44	267.66	356.88	446.10	535.32	624.55	713.77	802.99
20	89.04	178.08	267.12	356.16	445.20	534.24	623.28	712.32	801.36
30	88.86	177.71	266.57	355.43	444.25	533.14	622.00	710.86	799.72
40	88.67	177.35	266.02	354.69	443.37	532.04	620.72	709.39	798.06
50	88.49	176.98	265.47	353.95	442.44	530.93	619.42	707.91	796.40

TABLE 4A—*continued*

Angle °′	Apparent Slant Range								
	100	200	300	400	500	600	700	800	900
20 00	88.30	176.60	264.91	353.21	441.51	529.81	618.12	706.42	794.72
10	88.11	176.23	264.34	352.46	440.57	528.69	616.80	704.92	793.03
20	87.93	175.85	263.78	351.70	439.63	527.55	615.48	703.41	791.33
30	87.74	175.47	263.21	350.94	438.68	526.41	614.15	701.88	789.62
40	87.54	175.09	262.63	350.18	437.72	525.26	612.81	700.35	787.90
50	87.35	174.70	262.05	349.41	436.76	524.11	611.46	698.81	786.16
21 00	87.16	174.31	261.47	348.63	435.79	522.94	610.10	697.26	784.42
10	86.96	173.92	260.89	347.85	434.81	521.77	608.73	695.70	782.66
20	86.77	173.53	260.30	347.06	433.83	520.59	607.36	694.12	780.89
30	86.57	173.14	259.70	346.27	432.84	519.41	605.97	692.54	779.11
40	86.37	172.74	259.11	345.47	431.84	518.21	604.58	690.95	777.32
50	86.17	172.34	258.51	344.67	430.84	517.01	603.18	689.35	775.52
22 00	85.97	171.93	257.90	343.87	429.83	515.80	601.77	687.74	773.70
10	85.76	171.53	257.29	343.06	428.82	514.59	600.35	686.11	771.88
20	85.56	171.12	256.68	342.24	427.80	513.36	598.92	684.48	770.04
30	85.36	170.71	256.07	341.42	426.78	512.13	597.49	682.84	768.20
40	85.15	170.30	255.45	340.60	425.75	510.89	596.04	681.19	766.34
50	84.94	169.88	254.82	339.77	424.71	509.65	594.59	679.53	764.47
23 00	84.73	169.47	254.20	338.93	423.66	508.40	593.13	677.86	762.60
10	84.52	169.05	253.57	338.09	422.62	507.14	591.66	676.18	760.71
20	84.31	168.62	252.94	337.25	421.56	505.87	590.18	674.50	758.81
30	84.10	168.20	252.30	336.40	420.50	504.60	588.70	672.80	756.90
40	83.89	167.77	251.66	335.55	419.43	503.32	587.21	671.09	754.98
50	83.67	167.34	251.02	334.69	418.36	502.03	585.70	669.38	753.05
24 00	83.46	166.91	250.37	333.83	417.28	500.74	584.20	667.65	751.11
10	83.24	166.48	249.72	332.96	416.20	499.44	582.68	665.92	749.16
20	83.02	166.04	249.07	332.09	415.11	498.13	581.15	664.18	747.20
30	82.80	165.61	248.41	331.21	414.01	496.82	579.62	662.42	745.23
40	82.58	165.17	247.75	330.33	412.91	495.50	578.08	660.66	743.25
50	82.36	164.72	247.09	329.45	411.81	494.17	576.53	658.89	741.25
25 00	82.14	164.28	246.42	328.56	410.70	492.84	574.98	657.11	739.25
10	81.92	163.83	245.75	327.66	409.58	491.50	573.41	655.33	737.24
20	81.69	163.38	245.07	326.77	408.46	490.15	571.84	653.53	735.22
30	81.47	162.93	244.40	325.86	407.33	488.80	570.26	651.73	733.19
40	81.24	162.48	243.72	324.96	406.20	487.44	568.68	649.92	731.15
50	81.01	162.02	243.04	324.05	405.06	486.07	567.08	648.09	729.11
26 00	80.78	161.57	242.35	323.13	403.92	484.70	565.48	646.26	727.05
10	80.55	161.11	241.66	322.21	402.77	483.32	563.87	644.43	724.98
20	80.32	160.65	240.97	321.29	401.61	481.94	562.26	642.58	722.90
30	80.09	160.18	240.27	320.36	400.45	480.54	560.64	640.73	720.82
40	79.86	159.72	239.57	319.43	399.29	479.15	559.01	638.86	718.72
50	79.62	159.25	238.87	318.50	398.12	477.74	557.37	636.99	716.62
27 00	79.39	158.78	238.17	317.56	396.95	476.34	555.72	635.11	714.50
10	79.15	158.31	237.46	316.61	395.77	474.92	554.07	633.23	712.38
20	78.92	157.83	236.75	315.67	394.58	473.50	552.42	631.33	710.25
30	78.68	157.36	236.04	314.72	393.39	472.07	550.75	629.43	708.11
40	78.44	156.88	235.32	313.76	392.20	470.64	549.08	627.52	705.96
50	78.20	156.40	234.60	312.80	391.00	469.20	547.40	625.60	703.80
28 00	77.96	155.92	233.88	311.84	389.80	467.76	545.72	623.68	701.64
10	77.72	155.44	233.15	310.87	388.59	466.31	544.03	621.74	699.46
20	77.48	154.95	232.43	309.90	387.38	464.85	542.33	619.80	697.28
30	77.23	154.46	231.70	308.93	386.16	463.39	540.62	617.86	695.09
40	76.99	153.98	230.96	307.95	384.94	461.93	538.91	615.90	692.89
50	76.74	153.48	230.23	306.97	383.71	460.45	537.20	613.94	690.68
29 00	76.50	152.99	229.49	305.98	382.48	458.98	535.47	611.97	688.46
10	76.25	152.50	228.75	305.00	381.24	457.49	533.74	609.99	686.24
20	76.00	152.00	228.00	304.00	380.00	456.00	532.01	608.01	684.01
30	75.75	151.50	227.26	303.01	378.76	454.51	530.26	606.02	681.77
40	75.50	151.00	226.51	302.01	377.51	453.01	528.51	604.02	679.52
50	75.25	150.50	225.75	301.01	376.26	451.51	526.76	602.01	677.26

TABLE 4A—*continued*

Angle ° '	Apparent Slant Range								
	100	200	300	400	500	600	700	800	900
30 00	75.00	150.00	225.00	300.00	375.00	450.00	525.00	600.00	675.00
10	74.75	149.50	224.24	298.99	373.74	448.49	523.23	597.98	672.73
20	74.49	148.99	223.48	297.98	372.47	446.97	521.46	595.96	670.45
30	74.24	148.48	222.72	296.96	371.20	445.44	519.68	593.92	668.16
40	73.99	147.97	221.96	295.94	369.93	443.91	517.90	591.89	665.87
50	73.73	147.46	221.19	294.92	368.65	442.38	516.11	589.84	663.57
31 00	73.47	146.95	220.42	293.89	367.37	440.84	514.31	587.79	661.26
10	73.22	146.43	219.65	292.87	366.08	439.30	512.51	585.73	658.95
20	72.96	145.92	218.87	291.83	364.79	437.75	510.71	583.67	656.62
30	72.70	145.40	218.10	290.80	363.50	436.20	508.90	581.60	654.30
40	72.44	144.88	217.32	289.76	362.20	434.64	507.08	579.52	651.96
50	72.18	144.36	216.54	288.72	360.90	433.08	505.26	577.44	649.62
32 00	71.92	143.84	215.76	287.67	359.59	431.51	503.43	575.35	647.27
10	71.66	143.31	214.97	286.63	358.28	429.94	501.60	573.25	644.91
20	71.39	142.79	214.18	285.58	356.97	428.36	499.76	571.15	642.55
30	71.13	142.26	213.39	284.52	355.65	426.79	497.92	569.05	640.18
40	70.87	141.73	212.60	283.47	354.33	425.20	496.07	566.94	637.80
50	70.60	141.20	211.81	282.41	353.01	423.61	494.22	564.82	635.42
33 00	70.34	140.67	211.01	281.35	351.68	422.02	492.36	562.69	633.03
10	70.07	140.14	210.21	280.28	350.35	420.42	490.49	560.57	630.64
20	69.80	139.61	209.41	279.22	349.02	418.82	488.63	558.43	628.24
30	69.54	139.07	208.61	278.15	347.68	417.22	486.76	556.29	625.83
40	69.27	138.54	207.81	277.07	346.34	415.61	484.88	554.15	623.42
50	69.00	138.00	207.00	276.00	345.00	414.00	483.00	552.00	621.00
34 00	68.73	137.46	206.19	274.92	343.65	412.38	481.11	549.84	618.57
10	68.46	136.92	205.38	273.84	342.30	410.76	479.22	547.68	616.14
20	68.19	136.38	204.57	272.76	340.95	409.14	477.33	545.52	613.71
30	67.92	135.84	203.76	271.67	339.59	407.51	475.43	543.35	611.27
40	67.65	135.29	202.94	270.59	338.23	405.88	473.53	541.17	608.82
50	67.37	134.75	202.12	269.50	336.87	404.24	471.62	538.99	606.37
35 00	67.10	134.20	201.30	268.40	335.50	402.61	469.71	536.81	603.91
10	66.83	133.65	200.48	267.31	334.14	400.96	467.79	534.62	601.45
20	66.55	133.11	199.66	266.21	332.77	399.32	465.87	532.43	598.98
30	66.28	132.56	198.84	265.11	331.39	397.67	463.95	530.23	596.51
40	66.00	132.01	198.01	264.01	330.02	396.02	462.02	528.02	594.03
50	65.73	131.45	197.18	262.91	328.64	394.36	460.09	525.82	591.55
36 00	65.45	130.90	196.35	261.80	327.25	392.70	458.16	523.61	589.06
10	65.17	130.35	195.52	260.70	325.87	391.04	456.22	521.39	586.57
20	64.90	129.79	194.69	259.59	324.48	389.38	454.28	519.17	584.07
30	64.62	129.24	193.86	258.47	323.09	387.71	452.33	516.95	581.57
40	64.34	128.68	193.02	257.36	321.70	386.04	450.38	514.72	579.06
50	64.06	128.12	192.18	256.24	320.31	384.37	448.43	512.49	576.55
37 00	63.78	127.56	191.35	255.13	318.91	382.69	446.47	510.25	574.04
10	63.50	127.00	190.51	254.01	317.51	381.01	444.51	508.02	571.52
20	63.22	126.44	189.67	252.89	316.11	379.33	442.55	505.77	569.00
30	62.94	125.88	188.82	251.76	314.70	377.65	440.59	503.53	566.47
40	62.66	125.32	187.98	250.64	313.30	375.96	438.62	501.28	563.94
50	62.38	124.76	187.13	249.51	311.89	374.27	436.65	499.03	561.40
38 00	62.10	124.19	186.29	248.38	310.48	372.58	434.67	496.77	558.86
10	61.81	123.63	185.44	247.25	309.07	370.88	432.70	494.51	556.32
20	61.53	123.06	184.59	246.12	307.65	369.18	430.72	492.25	553.78
30	61.25	122.50	183.74	244.99	306.24	367.49	428.73	489.98	551.23
40	60.96	121.93	182.89	243.86	304.82	365.78	426.75	487.71	548.68
50	60.68	121.36	182.04	242.72	303.40	364.08	424.76	485.44	546.12
39 00	60.40	120.79	181.19	241.58	301.98	362.37	422.77	483.16	543.56
10	60.11	120.22	180.33	240.44	300.55	360.67	420.78	480.89	541.00
20	59.83	119.65	179.48	239.30	299.13	358.95	418.78	478.61	538.43
30	59.54	119.08	178.62	238.16	297.70	357.24	416.78	476.32	535.86
40	59.25	118.51	177.76	237.02	296.27	355.53	414.78	474.04	533.29
50	58.97	117.94	176.91	235.87	294.84	353.81	412.78	471.75	530.72

TABLE 4A—*continued*

Angle °,'	Apparent Slant Range								
	100	200	300	400	500	600	700	800	900
40 00	58.68	117.36	176.05	234.73	293.41	352.09	410.78	469.46	528.14
10	58.40	116.79	175.19	233.58	291.98	350.37	408.77	467.17	525.56
20	58.11	116.22	174.33	232.44	290.54	348.65	406.76	464.87	522.98
30	57.82	115.64	173.47	231.29	289.11	346.93	404.75	462.57	520.40
40	57.53	115.07	172.60	230.14	287.67	345.21	402.74	460.27	517.81
50	57.25	114.49	171.74	228.99	286.23	343 48	400.73	457.97	515.22
41 00	56.96	113.92	170.88	227.83	284.79	341.75	398.71	455.67	512.63
10	56.67	113.34	170.01	226.68	283.35	340.02	396.69	453.36	510.03
20	56.38	112.76	169.15	225.53	281.91	338.29	394.67	451.06	507.44
30	56.09	112.19	168.28	224.37	280.47	336.56	392.65	448.75	504.84
40	55.80	111.61	167.41	223.22	279.02	334.83	390.63	446.44	502.24
50	55.52	111.03	166.55	222.06	277.58	333.09	388.61	444.13	499.64
42 00	55.23	110.45	165.68	220.91	276.13	331.36	386.58	441.81	497.04
10	54.94	109.87	164.81	219.75	274.69	329.62	384.56	439.50	494.43
20	54.65	109.29	163.94	218.59	273.24	327.88	382.53	437.18	491.83
30	54.36	108.72	163.07	217.43	271.79	326.15	380.50	434.86	489.22
40	54.07	108.14	162.20	216.27	270.34	324.41	378.48	432.54	486.61
50	53.78	107.56	161.33	215.11	268.89	322.67	376.45	430.22	484.00
43 00	53.49	106.98	160.46	213.95	276.44	320.93	374.41	427.90	481.39
10	53.20	106.40	159.59	212.79	265.99	319.19	372.38	425.58	478.78
20	52.91	105.81	158.72	211.63	264.54	317.44	370.35	423.26	476.17
30	52.62	105.23	157.85	210.47	263.08	315.70	368.32	420.93	473.55
40	52.33	104.65	156.98	209.31	261.63	313.96	366.28	418.61	470.94
50	52.04	104.07	156.11	208.14	260.18	312.21	364.25	416.29	468.32
44 00	51.74	103.49	155.23	206.98	258.72	310.47	362.21	413.96	465.70
10	51.45	102.91	154.36	205.82	257.27	308.73	360.18	411.63	463.09
20	51.16	102.33	153.49	204.65	255.82	306.98	358.14	409.31	460.47
30	50.87	101.75	152.62	203.49	254.36	305.24	356.11	406.98	457.85
40	50.58	101.16	151.75	202.33	252.91	303.49	354.07	404.65	455.24
50	50.29	100.58	150.87	201.16	251.45	301.75	352.04	402.33	452.62
45 00	50.00	100.00	150.00	200.00	250.00	300.00	350.00	400.00	450.00

Example of use of the table:

At A the telescope is 1.62 m above the ground. A sight is taken to a staff held at B, with the telescope crosswire brought to the 1.620 mark on the staff.

> The upper stadia reads   2.880
> The lower stadia reads   0.410
> Apparent slant range is   247 m

B is higher than A, and the angle of elevation, to the nearest 10 minutes, is $11° 20'$.

From the table: at $11° 20'$ 200 m = 192.28
                                   40   = 38.46
                                    7   =   6.73
                                         ———————
                                          237.47

The true horizontal distance AB is 237.47 m.

TABLE 4B   True vertical differences from stadia intercepts

Angle ° '	Apparent Slant Range								
	100	200	300	400	500	600	700	800	900
00 00	00.00	00.00	00.00	00.00	00.00	00.00	00.00	00.00	00.00
10	00.29	00.58	00.87	01.16	01.45	01.75	02.04	02.33	02.62
20	00.58	01.16	01.75	02.33	02.91	03.49	04.07	04.65	05.24
30	00.87	01.75	02.62	03.49	04.36	05.24	06.11	06.98	07.85
40	01.16	02.33	03.49	04.65	05.82	06.98	08.14	09.31	10.47
50	01.45	02.91	04.36	05.82	07.27	08.73	10.18	11.63	13.09
01 00	01.74	03.49	05.23	06.98	08.72	10.47	12.21	13.96	15.70
10	02.04	04.07	06.11	08.14	10.18	12.21	14.25	16.29	18.32
20	02.33	04.65	06.98	09.31	11.63	13.96	16.28	18.61	20.94
30	02.62	05.23	07.85	10.47	13.08	15.70	18.32	20.93	23.55
40	02.91	05.81	08.72	11.63	14.54	17.44	20.35	23.26	26.17
50	03.20	06.40	09.59	12.79	15.99	19.19	22.38	25.58	28.78
02 00	03.49	06.98	10.46	13.95	17.44	20.93	24.41	27.90	31.39
10	03.78	07.56	11.33	15.11	18.89	22.67	26.45	30.22	34.00
20	04.07	08.14	12.20	16.27	20.34	24.41	28.48	32.54	36.61
30	04.36	08.72	13.07	17.43	21.79	26.15	30.50	34.86	39.22
40	04.65	09.29	13.94	18.59	23.24	27.88	32.53	37.18	41.83
50	04.94	09.87	14.81	19.75	24.69	29.62	34.56	39.50	44.43
03 00	05.23	10.45	15.68	20.91	26.13	31.36	36.58	41.81	47.04
10	05.52	11.03	16.55	22.06	27.58	33.09	38.61	44.13	49.64
20	05.80	11.61	17.41	23.22	29.02	34.83	40.63	46.44	52.24
30	06.09	12.19	18.28	24.37	30.47	36.56	42.65	48.75	54.84
40	06.38	12.76	19.15	25.53	31.91	38.29	44.67	51.06	57.44
50	06.67	13.34	20.01	26.68	33.35	40.02	46.69	53.36	60.03
04 00	06.96	13.92	20.88	27.83	34.79	41.75	48.71	55.67	62.63
10	07.25	14.49	21.74	28.99	36.23	43.48	50.73	57.97	65.22
20	07.53	15.07	22.60	30.14	37.67	45.21	52.74	60.27	67.81
30	07.82	15.64	23.47	31.29	39.11	46.93	54.75	62.57	70.40
40	08.11	16.22	24.33	32.44	40.54	48.65	56.76	64.87	72.98
50	08.40	16.79	25.19	33.58	41.98	50.37	58.77	67.17	75.56
05 00	08.68	17.36	26.05	34.73	43.41	52.09	60.78	69.46	78.14
10	08.97	17.94	26.91	35.87	44.84	53.81	62.78	71.75	80.72
20	09.25	18.51	27.76	37.02	46.27	55.53	64.78	74.04	83.29
30	09.54	19.08	28.62	38.16	47.70	57.24	66.78	76.32	85.86
40	09.83	19.65	29.48	39.30	49.13	58.95	68.78	78.61	88.43
50	10.11	20.22	30.33	40.44	50.55	60.67	70.78	80.89	91.00
06 00	10.40	20.79	31.19	41.58	51.98	62.37	72.77	83.16	93.56
10	10.68	21.36	32.04	42.72	53.40	64.08	74.76	85.44	96.12
20	10.96	21.93	32.89	43.86	54.82	65.78	76.75	87.71	98.68
30	11.25	22.50	33.74	44.99	56.24	67.49	78.73	89.98	101.23
40	11.53	23.06	34.59	46.12	57.65	69.18	80.72	92.25	103.78
50	11.81	23.63	35.44	47.25	59.07	70.88	82.70	94.51	106.32
07 00	12.10	24.19	36.29	48.38	60.48	72.58	84.67	96.77	108.86
10	12.38	24.76	37.13	49.51	61.89	74.27	86.65	99.02	111.40
20	12.66	25.32	37.98	50.64	63.30	75.96	88.62	101.28	113.94
30	12.94	25.88	38.82	51.76	64.70	77.65	90.59	103.53	116.47
40	13.22	26.44	39.67	52.89	66.11	79.33	92.55	105.77	119.00
50	13.50	27.00	40.51	54.01	67.51	81.01	94.51	108.02	121.52
08 00	13.78	27.56	41.35	55.13	68.91	82.69	96.47	110.25	124.04
10	14.06	28.12	42.18	56.24	70.31	84.37	98.43	112.49	126.55
20	14.34	28.68	43.02	57.36	71.70	86.04	100.38	114.72	129.06
30	14.62	29.24	43.86	58.47	73.09	87.71	102.33	116.95	131.57
40	14.90	29.79	44.69	59.59	74.48	89.38	104.28	119.17	134.07
50	15.17	30.35	45.52	60.70	75.87	91.04	106.22	121.39	136.57
09 00	15.45	30.90	46.35	61.80	77.25	92.71	108.16	123.61	139.06
10	15.73	31.45	47.18	62.91	78.64	94.36	110.09	125.82	141.55
20	16.00	32.01	48.01	64.01	80.02	96.02	112.02	128.02	144.03
30	16.28	32.56	48.84	65.11	81.39	97.67	113.95	130.23	146.51
40	16.55	33.11	49.66	66.21	82.77	99.32	115.87	132.43	148.98
50	16.83	33.65	50.48	67.31	84.14	100.96	117.79	134.62	151.45

TABLE 4B—*continued*

Angle ° '	Apparent Slant Range								
	100	200	300	400	500	600	700	800	900
10 00	17.10	34.20	51.30	68.40	85.50	102.61	119.71	136.81	153.91
10	17.37	34.75	52.12	69.50	86.87	104.24	121.62	138.99	156.37
20	17.65	35.29	52.94	70.59	88.23	105.88	123.53	141.17	158.82
30	17.92	35.84	53.76	71.67	89.59	107.51	125.43	143.35	161.27
40	18.19	36.38	54.57	72.76	90.95	109.14	127.33	145.52	163.71
50	18.46	36.92	55.38	73.84	92.30	110.76	129.22	147.68	166.14
11 00	18.73	37.46	56.19	74.92	93.65	112.38	131.11	149.84	168.57
10	19.00	38.00	57.00	76.00	95.00	114.00	133.00	152.00	171.00
20	19.27	38.54	57.81	77.07	96.34	115.61	134.88	154.15	173.42
30	19.54	39.07	58.61	78.15	97.68	117.22	136.76	156.29	175.83
40	19.80	39.61	59.41	79.22	99.02	118.82	138.63	158.43	178.24
50	20.07	40.14	60.21	80.28	100.35	120.42	140.49	160.57	180.64
12 00	20.34	40.67	61.01	81.35	101.68	122.02	142.36	162.69	183.03
10	20.60	41.20	61.81	82.41	103.01	123.61	144.22	164.82	185.42
20	20.87	41.73	62.60	83.47	104.33	125.20	146.07	166.94	187.80
30	21.13	42.26	63.39	84.52	105.65	126.79	147.92	169.05	190.18
40	21.39	42.79	64.18	85.58	106.97	128.36	149.76	171.15	192.55
50	21.66	43.31	64.97	86.63	108.28	129.94	151.60	173.25	194.91
13 00	21.92	43.84	65.76	87.67	109.59	131.51	153.43	175.35	197.27
10	22.18	44.36	66.54	88.72	110.90	133.08	155.26	177.44	199.62
20	22.44	44.88	67.32	89.76	112.20	134.64	157.08	179.52	201.96
30	22.70	45.40	68.10	90.80	113.50	136.20	158.90	181.60	204.30
40	22.96	45.92	68.87	91.83	114.79	137.75	160.71	183.67	206.62
50	23.22	46.43	69.65	92.87	116.08	139.30	162.51	185.73	208.95
14 00	23.47	46.95	70.42	93.89	117.37	140.84	164.31	187.79	211.26
10	23.73	47.46	71.19	94.92	118.65	142.38	166.11	189.84	213.57
20	23.99	47.97	71.96	95.94	119.93	143.91	167.90	191.89	215.87
30	24.24	48.48	72.72	96.96	121.20	145.44	169.68	193.92	218.16
40	24.49	48.99	73.48	97.98	122.47	146.97	171.46	195.96	220.45
50	24.75	49.50	74.24	98.99	123.74	148.49	173.23	197.98	222.73
15 00	25.00	50.00	75.00	100.00	125.00	150.00	175.00	200.00	225.00
10	25.25	50.50	75.75	101.01	126.26	151.51	176.76	202.01	227.26
20	25.50	51.00	76.51	102.01	127.51	153.01	178.51	204.02	229.52
30	25.75	51.50	77.26	103.01	128.76	154.51	180.26	206.01	231.77
40	26.00	52.00	78.00	104.00	130.00	156.00	182.01	208.01	234.01
50	26.25	52.50	78.75	105.00	131.24	157.49	183.74	209.99	236.24
16 00	26.50	52.99	79.49	105.98	132.48	158.98	185.47	211.97	238.46
10	26.74	53.48	80.23	106.97	133.71	160.45	187.20	213.94	240.68
20	26.99	53.98	80.96	107.95	134.94	161.93	188.91	215.90	242.89
30	27.23	54.46	81.70	108.93	136.16	163.39	190.62	217.86	245.09
40	27.48	54.95	82.43	109.90	137.38	164.85	192.33	219.80	247.28
50	27.72	55.44	83.15	110.87	138.59	166.31	194.03	221.74	249.46
17 00	27.96	55.92	83.88	111.84	139.80	167.76	195.72	223.68	251.64
10	28.20	56.40	84.60	112.80	141.00	169.20	197.40	225.60	253.80
20	28.44	56.88	85.32	113.76	142.20	170.64	199.08	227.52	255.96
30	28.68	57.36	86.04	114.72	143.39	172.07	200.75	229.43	258.11
40	28.92	57.83	86.75	115.67	144.58	173.50	202.42	231.33	260.25
50	29.15	58.31	87.46	116.61	145.77	174.92	204.07	233.23	262.38
18 00	29.39	58.78	88.17	117.56	146.95	176.34	205.72	235.11	264.50
10	29.62	59.25	88.87	118.50	148.12	177.74	207.37	236.99	266.62
20	29.86	59.72	89.57	119.43	149.29	179.15	209.01	238.86	268.72
30	30.09	60.18	90.27	120.36	150.45	180.54	210.63	240.73	270.82
40	30.32	60.65	90.97	121.29	151.61	181.94	212.26	242.58	272.90
50	30.55	61.11	91.66	122.21	152.77	183.32	213.87	244.43	274.98
19 00	30.78	61.57	92.35	123.13	153.92	184.70	215.48	246.26	277.05
10	31.01	62.02	93.04	124.05	155.06	186.07	217.08	248.09	279.11
20	31.24	62.48	93.72	124.96	156.20	187.44	218.68	249.92	281.15
30	31.47	62.93	94.40	125.86	157.33	188.80	220.26	251.73	283.19
40	31.69	63.38	95.07	126.77	158.46	190.15	221.84	253.53	285.22
50	31.92	63.83	95.75	127.66	159.58	191.50	223.41	255.33	287.24

TABLE 4B—*continued*

Angle ° ′	Apparent Slant Range								
	100	200	300	400	500	600	700	800	900
20 00	32.14	64.28	96.42	128.56	160.70	192.84	224.98	257.11	289.25
10	32.36	64.72	97.08	129.45	161.81	194.17	226.53	258.89	291.25
20	32.58	65.17	97.75	130.33	162.91	195.50	228.08	260.66	293.25
30	32.80	65.61	98.41	131.21	164.01	196.82	229.62	262.42	295.23
40	33.02	66.04	99.07	132.09	165.11	198.13	231.15	264.18	297.20
50	33.24	66.48	99.72	132.96	166.20	199.44	232.68	265.92	299.16
21 00	33.46	66.91	100.37	133.83	167.28	200.74	234.20	267.65	301.11
10	33.67	67.34	101.02	134.69	168.36	202.03	235.70	269.38	303.05
20	33.89	67.77	101.66	135.55	169.43	203.32	237.21	271.09	304.98
30	34.10	68.20	102.30	136.40	170.50	204.60	238.70	272.80	306.90
40	34.31	68.62	102.94	137.25	171.56	205.87	240.18	274.50	308.81
50	34.52	69.05	103.57	138.09	172.62	207.14	241.66	276.18	310.71
22 00	34.73	69.47	104.20	138.93	173.66	208.40	243.13	277.86	312.60
10	34.94	69.88	104.82	139.77	174.71	209.65	244.59	279.53	314.47
20	35.15	70.30	105.45	140.60	175.75	210.89	246.04	281.19	316.34
30	35.36	70.71	106.07	141.42	176.78	212.13	247.49	282.84	318.20
40	35.56	71.12	106.68	142.24	177.80	213.36	248.92	284.48	320.06
50	35.76	71.53	107.29	143.06	178.82	214.59	250.35	286.11	321.88
23 00	35.97	71.93	107.90	143.87	179.83	215.80	251.77	287.74	323.70
10	36.17	72.34	108.51	144.67	180.84	217.01	253.18	289.35	325.52
20	36.37	72.74	109.11	145.47	181.84	218.21	254.58	290.95	327.32
30	36.57	73.14	109.70	146.27	182.84	219.41	255.97	292.54	329.11
40	36.77	73.53	110.30	147.06	183.83	220.59	257.36	294.12	330.89
50	36.96	73.92	110.89	147.85	184.81	221.77	258.73	295.70	332.66
24 00	37.16	74.31	111.47	148.63	185.79	222.94	260.10	297.26	334.41
10	37.35	74.70	112.05	149.40	186.76	224.11	261.46	298.81	336.16
20	37.54	75.09	112.63	150.18	187.72	225.26	262.81	300.35	337.90
30	37.74	75.47	113.21	150.94	188.68	226.41	264.15	301.88	339.62
40	37.93	75.85	113.78	151.70	189.63	227.55	265.48	303.41	341.33
50	38.11	76.23	114.34	152.46	190.57	228.69	266.80	304.92	343.03
25 00	38.30	76.60	114.91	153.21	191.51	229.81	268.12	306.42	344.72
10	38.49	76.98	115.47	153.95	192.44	230.93	269.42	307.91	346.40
20	38.67	77.35	116.02	154.69	193.37	232.04	270.71	309.39	348.06
30	38.86	77.71	116.57	155.43	194.29	233.14	272.00	310.86	349.72
40	39.04	78.08	117.12	156.16	195.20	234.24	273.28	312.32	351.36
50	39.22	78.44	117.66	156.88	196.10	235.32	274.54	313.77	352.99
26 00	39.40	78.80	118.20	157.60	197.00	236.40	275.80	315.20	354.60
10	39.58	79.16	118.74	158.32	197.89	237.47	277.05	316.63	356.21
20	39.76	79.51	119.27	159.02	198.78	238.54	278.29	318.05	357.80
30	39.93	79.86	119.80	159.73	199.66	239.59	279.52	319.45	359.39
40	40.11	80.21	120.32	160.42	200.53	240.64	280.74	320.85	360.96
50	40.28	80.56	120.84	161.12	201.40	241.67	281.95	322.23	362.51
27 00	40.45	80.90	121.35	161.80	202.25	242.71	283.16	323.61	364.06
10	40.62	81.24	121.86	162.48	203.11	243.73	284.35	324.97	365.59
20	40.79	81.58	122.37	163.16	203.95	244.74	285.53	326.32	367.11
30	40.96	81.92	122.87	163.83	204.79	245.75	286.70	327.66	368.62
40	41.12	82.25	123.37	164.49	205.62	246.74	287.87	328.99	370.11
50	41.29	82.58	123.87	165.15	206.44	247.73	289.02	330.31	371.60
28 00	41.45	82.90	124.36	165.81	207.26	248.71	290.16	331.61	373.07
10	41.61	83.23	124.84	166.46	208.07	249.68	291.30	332.91	374.52
20	41.77	83.55	125.32	167.10	208.87	250.65	292.42	334.19	375.97
30	41.93	83.87	125.80	167.73	209.67	251.60	293.53	335.47	377.40
40	42.09	84.18	126.27	168.36	210.46	252.55	294.64	336.73	378.82
50	42.25	84.50	126.74	168.99	211.24	253.49	295.73	337.98	380.23
29 00	42.40	84.80	127.21	169.61	212.01	254.41	296.82	339.22	381.62
10	42.56	85.11	127.67	170.22	212.78	255.33	297.89	340.45	383.00
20	42.71	85.42	128.12	170.83	213.54	256.25	298.95	341.66	384.37
30	42.86	85.72	128.58	171.43	214.29	257.15	300.01	342.87	385.72
40	43.01	86.01	129.02	172.03	215.04	258.04	301.05	344.06	387.07
50	43.16	86.31	129.47	172.62	215.78	258.93	302.09	345.24	388.40

TABLE 4B—*continued*

Angle ° '	Apparent Slant Range								
	100	200	300	400	500	600	700	800	900
30 00	43.30	86.60	129.90	173.21	216.51	259.81	303.11	346.41	389.71
10	43.45	86.89	130.34	173.78	217.23	260.68	304.12	347.57	391.01
20	43.59	87.18	130.77	174.36	217.95	261.53	305.12	348.71	392.30
30	43.73	87.46	131.19	174.92	218.65	262.39	306.12	349.85	393.58
40	43.87	87.74	131.61	175.48	219.36	263.23	307.10	350.97	394.84
50	44.01	88.02	132.03	176.04	220.05	264.06	308.07	352.08	396.09
31 00	44.15	88.29	132.44	176.59	220.74	264.88	309.03	353.18	397.33
10	44.28	88.57	132.85	177.13	221.42	265.70	309.98	354.27	398.55
20	44.42	88.83	133.25	177.07	222.09	266.50	310.92	355.34	399.76
30	44.55	89.10	133.65	178.20	222.75	267.30	311.85	356.40	400.95
40	44.68	89.36	134.04	178.73	223.41	268.09	312.77	357.45	402.13
50	44.81	89.62	134.43	179.25	224.06	268.87	313.68	358.49	403.30
32 00	44.94	89.88	134.82	179.76	224.70	269.64	314.58	359.52	404.46
10	45.07	90.13	135.20	180.27	225.33	270.40	315.46	360.53	405.60
20	45.19	90.38	135.57	180.77	225.96	271.15	316.34	361.53	406.72
30	45.32	90.63	135.95	181.26	226.58	271.89	317.21	362.52	407.84
40	45.44	90.88	136.31	181.75	227.19	272.63	318.06	363.50	408.94
50	45.56	91.12	136.67	182.23	227.79	273.35	318.91	364.47	410.02
33 00	45.68	91.35	137.03	182.71	228.39	274.06	319.74	365.42	411.10
10	45.79	91.59	137.38	183.18	228.97	274.77	320.56	366.36	412.15
20	45.91	91.82	137.73	183.64	229.55	275.46	321.38	367.29	413.20
30	46.03	92.05	138.08	184.10	230.13	276.15	322.18	368.20	414.23
40	46.14	92.28	138.41	184.55	230.69	276.83	322.97	369.10	415.24
50	46.25	92.50	138.75	185.00	231.25	277.50	323.75	370.00	416.24
34 00	46.36	92.72	139.08	185.44	231.80	278.16	324.51	370.87	417.23
10	46.47	92.93	139.40	185.87	232.34	278.80	325.27	371.74	418.21
20	46.57	93.15	139.72	186.30	232.87	279.44	326.02	372.59	419.17
30	46.68	93.36	140.04	186.72	233.40	280.07	326.75	373.43	420.11
40	46.78	93.56	140.35	187.13	233.91	280.69	327.48	374.26	421.04
50	46.88	93.77	140.65	187.54	234.42	281.31	328.19	375.07	421.96
35 00	46.98	93.97	140.95	187.94	234.92	281.91	328.89	375.88	422.86
10	47.08	94.17	141.25	188.33	235.42	282.50	329.58	376.67	423.75
20	47.18	94.36	141.54	188.72	235.90	283.08	330.26	377.44	424.62
30	47.28	94.55	141.83	189.10	236.38	283.66	330.93	378.21	425.48
40	47.37	94.74	142.11	189.48	236.85	284.22	331.59	378.96	426.33
50	47.46	94.92	142.39	189.85	237.31	284.77	332.23	379.70	427.16
36 00	47.55	95.11	142.66	190.21	237.76	285.32	332.87	380.42	427.98
10	47.64	95.28	142.93	190.57	238.21	285.85	333.49	381.14	428.78
20	47.73	95.46	143.19	190.92	238.65	286.38	334.11	381.83	429.56
30	47.82	95.63	143.45	191.26	239.08	286.89	334.71	382.52	430.34
40	47.90	95.80	143.70	191.60	239.50	287.40	335.30	383.20	431.09
50	47.98	95.96	143.95	191.93	239.91	287.89	335.87	383.86	431.84
37 00	48.06	96.13	144.19	192.25	240.32	288.38	336.44	384.50	432.57
10	48.14	96.28	144.43	192.57	240.71	288.85	337.00	385.14	433.28
20	48.22	96.44	144.66	192.88	241.10	289.32	337.54	385.76	433.98
30	48.30	96.59	144.89	193.19	241.48	289.78	338.07	386.37	434.67
40	48.37	96.74	145.11	193.48	241.85	290.22	338.59	386.97	435.34
50	48.44	96.89	145.33	193.77	242.22	290.66	339.10	387.55	435.99
38 00	48.51	97.03	145.54	194.06	242.57	291.09	339.60	388.12	436.63
10	48.58	97.17	145.75	194.34	242.92	291.51	340.09	388.67	437.26
20	48.65	97.30	145.96	194.61	243.26	291.91	340.57	389.22	437.87
30	48.72	97.44	146.16	194.87	243.59	292.31	341.03	389.75	438.47
40	48.78	97.57	146.35	195.13	243.92	292.70	341.48	390.26	439.05
50	48.85	97.69	146.54	195.38	244.23	293.08	341.92	390.77	439.61
39 00	48.91	97.81	146.72	195.63	244.54	293.44	342.35	391.26	440.17
10	48.97	97.93	146.90	195.87	244.84	293.80	342.77	391.74	440.70
20	49.03	98.05	147.08	196.10	245.13	294.15	343.18	392.20	441.22
30	49.08	98.16	147.24	196.33	245.41	294.49	343.57	392.65	441.73
40	49.14	98.27	147.41	196.54	245.68	294.82	343.95	393.09	442.22
50	49.19	98.38	147.57	196.76	245.95	295.13	344.32	393.51	442.70

TABLE 4B—*continued*

Angle ° '	Apparent Slant Range								
	100	200	300	400	500	600	700	800	900
40 00	49.24	98.48	147.72	196.96	246.20	295.44	344.68	393.92	443.16
10	49.29	98.58	147.87	197.16	246.45	295.74	345.03	394.32	443.61
20	49.34	98.68	148.01	197.35	246.69	296.03	345.37	394.70	444.04
30	49.38	98.77	148.15	197.54	246.92	296.31	345.69	395.08	444.46
40	49.43	98.86	148.29	197.72	247.15	296.57	346.00	395.43	444.86
50	49.47	98.94	148.42	197.89	247.36	296.83	346.30	395.78	445.25
41 00	49.51	99.03	148.54	198.05	247.57	297.08	346.59	396.11	445.62
10	49.55	99.11	148.66	198.21	247.77	297.32	346.87	396.42	445.98
20	49.59	99.18	148.77	198.36	247.96	297.55	347.14	396.73	446.32
30	49.63	99.25	148.88	198.51	248.14	297.76	347.39	397.02	446.65
40	49.66	99.32	148.99	198.65	248.31	297.97	347.63	397.30	446.96
50	49.69	99.39	149.08	198.78	248.47	298.17	347.86	397.56	447.25
42 00	49.73	99.45	149.18	198.90	248.63	298.36	348.08	397.81	447.53
10	49.76	99.51	149.27	199.02	248.78	298.53	348.29	398.05	447.80
20	49.78	99.57	149.35	199.13	248.92	298.70	348.48	398.27	448.05
30	49.81	99.62	149.43	199.24	249.05	298.86	348.67	398.48	448.29
40	49.83	99.67	149.50	199.34	249.17	299.01	348.84	398.67	448.51
50	49.86	99.71	149.57	199.43	249.29	299.14	349.00	398.86	448.71
43 00	49.88	99.76	149.63	199.51	249.39	299.27	349.15	399.03	448.90
10	49.90	99.80	149.69	199.59	249.49	299.39	349.28	399.18	449.08
20	49.92	99.83	149.75	199.66	249.58	299.49	349.41	399.32	449.24
30	49.93	99.86	149.79	199.73	249.66	299.59	349.52	399.45	449.38
40	49.95	99.89	149.84	199.78	249.73	299.68	349.62	399.57	449.51
50	49.96	99.92	149.88	199.83	249.79	299.75	349.71	399.67	449.63
44 00	49.97	99.94	149.91	199.88	249.85	299.82	349.79	399.76	449.73
10	49.98	99.96	149.94	199.92	249.89	299.87	349.85	399.83	449.81
20	49.99	99.97	149.96	199.95	249.93	299.92	349.91	399.89	449.88
30	49.99	99.98	149.98	199.97	249.96	299.95	349.95	399.94	449.93
40	50.00	99.99	149.99	199.99	249.98	299.98	349.98	399.97	449.97
50	50.00	100.00	150.00	200.00	250.00	299.99	349.99	399.99	449.99
45 00	50.00	100.00	150.00	200.00	250.00	300.00	350.00	400.00	450.00

Example of use of the table:

At A the telescope is 1.62 m above the ground. A sight is taken to a staff held at B, with the telescope crosswire brought to the 1.620 mark on the staff.

The upper stadia reads	2.880
The lower stadia reads	0.410
Apparent slant range is	247 m

B is higher than A, and the angle of elevation, to the nearest 10 minutes, is 11° 20'.

From the table: at 11° 20'   200 m = 38.54
                              40 = 7.71
                              7   = 1.35
           Difference in altitude = 47.60

If the altitude of A is 167.83 m, the altitude of B is 215.43 m.

TABLE 5   Natural tangents

	Minutes					
Degrees	00	10	20	30	40	50
00	.0000	.0029	.0058	.0087	.0116	.0145
01	.0175	.0204	.0233	.0262	.0291	.0320
02	.0349	.0378	.0407	.0437	.0466	.0495
03	.0524	.0553	.0582	.0612	.0641	.0670
04	.0699	.0729	.0758	.0787	.0816	.0846
05	.0875	.0904	.0934	.0963	.0992	.1022
06	.1051	.1080	.1110	.1139	.1169	.1198
07	.1228	.1257	.1287	.1317	.1346	.1376
08	.1405	.1435	.1465	.1495	.1524	.1554
09	.1584	.1614	.1644	.1673	.1703	.1733
10	.1763	.1793	.1823	.1853	.1883	.1914
11	.1944	.1974	.2004	.2035	.2065	.2095
12	.2126	.2156	.2186	.2217	.2247	.2278
13	.2309	.2339	.2370	.2401	.2432	.2462
14	.2493	.2524	.2555	.2586	.2617	.2648
15	.2679	.2711	.2742	.2773	.2805	.2836
16	.2867	.2899	.2931	.2962	.2994	.3026
17	.3057	.3089	.3121	.3153	.3185	.3217
18	.3249	.3281	.3314	.3346	.3378	.3411
19	.3443	.3476	.3508	.3541	.3574	.3607
20	.3640	.3673	.3706	.3739	.3772	.3805
21	.3839	.3872	.3906	.3939	.3973	.4006
22	.4040	.4074	.4108	.4142	.4176	.4210
23	.4245	.4279	.4314	.4348	.4383	.4417
24	.4452	.4487	.4522	.4557	.4592	.4628
25	.4663	.4699	.4734	.4770	.4806	.4841
26	.4877	.4913	.4950	.4986	.5022	.5059
27	.5095	.5132	.5169	.5206	.5243	.5280
28	.5317	.5354	.5392	.5430	.5467	.5505
29	.5543	.5581	.5619	.5658	.5696	.5735
30	.5774	.5812	.5851	.5890	.5930	.5969
31	.6009	.6048	.6088	.6128	.6168	.6208
32	.6249	.6289	.6330	.6371	.6412	.6453
33	.6494	.6536	.6577	.6619	.6661	.6703
34	.6745	.6787	.6830	.6873	.6916	.6959
35	.7002	.7046	.7089	.7133	.7177	.7221
36	.7265	.7310	.7355	.7400	.7445	.7490
37	.7536	.7581	.7627	.7673	.7720	.7766
38	.7813	.7860	.7907	.7954	.8002	.8050
39	.8098	.8146	.8195	.8243	.8292	.8342

TABLE 6   Checking table for traverses

									Eastings				
Bearing (degrees)									Distance (units of 10)				
+	+	−	−	1	2	3	4	5	6	7	8	9
00	180	180	360	0.00	0.00	0.00	0.00	0.00	0.00	0.00	0.00	0.00
01	179	181	359	0.17	0.35	0.52	0.70	0.87	1.05	1.22	1.40	1.57
02	178	182	358	0.35	0.70	1.05	1.40	1.75	2.09	2.44	2.79	3.14
03	177	183	357	0.52	1.05	1.57	2.09	2.62	3.14	3.66	4.19	4.71
04	176	184	356	0.70	1.40	2.09	2.79	3.49	4.19	4.88	5.58	6.28
05	175	185	355	0.87	1.74	2.61	3.49	4.36	5.23	6.10	6.97	7.84
06	174	186	354	1.05	2.09	3.14	4.18	5.23	6.27	7.32	8.36	9.41
07	173	187	353	1.22	2.44	3.66	4.88	6.09	7.31	8.53	9.75	10.97
08	172	188	352	1.39	2.78	4.18	5.57	6.96	8.35	9.74	11.13	12.53
09	171	189	351	1.56	3.13	4.69	6.26	7.82	9.39	10.95	12.51	14.08
10	170	190	350	1.74	3.47	5.21	6.95	8.68	10.42	12.16	13.89	15.63
11	169	191	349	1.91	3.82	5.72	7.63	9.54	11.45	13.36	15.26	17.18
12	168	192	348	2.08	4.16	6.24	8.32	10.40	12.48	14.55	16.63	18.71
13	167	193	347	2.25	4.50	6.75	9.00	11.25	13.50	15.75	18.00	20.25
14	166	194	346	2.42	4.84	7.26	9.68	12.10	14.52	16.93	19.35	21.77
15	165	195	345	2.59	5.18	7.76	10.35	12.94	15.53	18.12	20.71	23.29
16	164	196	344	2.76	5.51	8.27	11.03	13.78	16.54	19.29	22.05	24.81
17	163	197	343	2.92	5.85	8.77	11.69	14.62	17.54	20.47	23.39	26.31
18	162	198	342	3.09	6.18	9.27	12.36	15.45	18.54	21.63	24.72	27.81
19	161	199	341	3.26	6.51	9.77	13.02	16.28	19.53	22.79	26.05	29.30
20	160	200	340	3.42	6.84	10.26	13.68	17.10	20.52	23.94	27.36	30.78
21	159	201	339	3.58	7.17	10.75	14.33	17.92	21.50	25.09	28.67	32.25
22	158	202	338	3.75	7.49	11.24	14.98	18.73	22.48	26.22	29.97	33.71
23	157	203	337	3.91	7.81	11.72	15.63	19.54	23.44	27.35	31.26	35.17
24	156	204	336	4.07	8.13	12.20	16.27	20.34	24.40	28.47	32.54	36.61
25	155	205	335	4.23	8.45	12.68	16.90	21.13	25.36	29.58	33.81	38.04
26	154	206	334	4.38	8.77	13.15	17.53	21.92	26.30	30.69	35.07	39.45
27	153	207	333	4.54	9.08	13.62	18.16	22.70	27.24	31.78	36.32	40.86
28	152	208	332	4.69	9.39	14.08	18.78	23.47	28.17	32.86	37.56	42.25
29	151	209	331	4.85	9.70	14.54	19.39	24.24	29.09	33.94	38.78	43.63
30	150	210	330	5.00	10.00	15.00	20.00	25.00	30.00	35.00	40.00	45.00
31	149	211	329	5.15	10.30	15.45	20.60	25.75	30.90	36.05	41.20	46.35
32	148	212	328	5.30	10.60	15.90	21.20	26.50	31.80	37.09	42.39	47.69
33	147	213	327	5.45	10.89	16.34	21.79	27.23	32.68	38.12	43.57	49.02
34	146	214	326	5.59	11.18	16.78	22.37	27.96	33.55	39.14	44.74	50.33
35	145	215	325	5.74	11.47	17.21	22.94	28.68	34.41	40.15	45.89	51.62
36	144	216	324	5.88	11.76	17.63	23.51	29.39	35.27	41.14	47.02	52.90
37	143	217	323	6.02	12.04	18.05	24.07	30.09	36.11	42.13	48.15	54.16
38	142	218	322	6.16	12.31	18.47	24.63	30.78	36.94	43.10	49.25	55.41
39	141	219	321	6.29	12.59	18.88	25.17	31.47	37.76	44.05	50.35	56.64
40	140	220	320	6.43	12.86	19.28	25.71	32.14	38.57	45.00	51.42	57.85
41	139	221	319	6.56	13.12	19.68	26.24	32.80	39.36	45.92	52.48	59.05
42	138	222	318	6.69	13.38	20.07	26.77	33.46	40.15	46.84	53.53	60.22
43	137	223	317	6.82	13.64	20.46	27.28	34.10	40.92	47.74	54.56	61.38
44	136	224	316	6.95	13.89	20.84	27.79	34.73	41.68	48.63	55.57	62.52
45	135	225	315	7.07	14.14	21.21	28.28	35.36	42.43	49.50	56.57	63.64
46	134	226	314	7.19	14.39	21.58	28.77	35.97	43.16	50.35	57.55	64.74
47	133	227	313	7.31	14.63	21.94	29.25	36.57	43.88	51.19	58.51	65.82
48	132	228	312	7.43	14.86	22.29	29.73	37.16	44.59	52.02	59.45	66.88
49	131	229	311	7.55	15.09	22.64	30.19	37.74	45.28	52.83	60.38	67.92
50	130	230	310	7.66	15.32	22.98	30.64	38.30	45.96	53.62	61.28	68.94
51	129	231	309	7.77	15.54	23.31	31.09	38.86	46.63	54.40	62.17	69.94
52	128	232	308	7.88	15.76	23.64	31.52	39.40	47.28	55.16	63.04	70.92
53	127	233	307	7.99	15.97	23.96	31.95	39.93	47.92	55.90	63.89	71.88
54	126	234	306	8.09	16.18	24.27	32.36	40.45	48.54	56.63	64.72	72.81
55	125	235	305	8.19	16.38	24.57	32.77	40.96	49.15	57.34	65.53	73.72
56	124	236	304	8.29	16.58	24.87	33.16	41.45	49.74	58.03	66.32	74.61
57	123	237	303	8.39	16.77	25.16	33.55	41.93	50.32	58.71	67.09	75.48
58	122	238	302	8.48	16.96	25.44	33.92	42.40	50.88	59.36	67.84	76.32
59	121	239	301	8.57	17.14	25.72	34.29	42.86	51.43	60.00	68.57	77.15

TABLE 6—*continued*

				Eastings								
Bearings (degrees)				Distance (units of 10)								
+	+	−	−	1	2	3	4	5	6	7	8	9
60	120	240	300	8.66	17.32	25.98	34.64	43.30	51.96	60.62	69.28	77.94
61	119	241	299	8.75	17.49	26.24	34.98	43.73	52.48	61.22	69.97	78.72
62	118	242	298	8.83	17.66	26.49	35.32	44.15	52.98	61.81	70.64	79.47
63	117	243	297	8.91	17.82	26.73	35.64	44.55	53.46	62.37	71.28	80.19
64	116	244	296	8.99	17.98	26.96	35.95	44.94	53.93	62.92	71.90	80.89
65	115	245	295	9.06	18.13	27.19	36.25	45.32	54.38	63.44	72.50	81.57
66	114	246	294	9.14	18.27	27.41	36.54	45.68	54.81	63.95	73.08	82.22
67	113	247	293	9.21	18.41	27.62	36.82	46.03	55.23	64.44	73.64	82.85
68	112	248	292	9.27	18.54	27.82	37.09	46.36	55.63	64.90	74.17	83.45
69	111	249	291	9.34	18.67	28.01	37.34	46.68	56.01	65.35	74.69	84.02
70	110	250	290	9.40	18.79	28.19	37.59	46.98	56.38	65.78	75.18	84.57
71	109	251	289	9.46	18.91	28.37	37.82	47.28	56.73	66.19	75.64	85.10
72	108	252	288	9.51	19.02	28.53	38.04	47.55	57.06	66.57	76.08	85.60
73	107	253	287	9.56	19.13	28.69	38.25	47.82	57.38	66.94	76.50	86.07
74	106	254	286	9.61	19.23	28.84	38.45	48.06	57.68	67.29	76.90	86.51
75	105	255	285	9.66	19.32	28.98	38.64	48.30	57.96	67.61	77.27	86.93
76	104	256	284	9.70	19.41	29.11	38.81	48.51	58.22	67.92	77.62	87.33
77	103	257	283	9.74	19.49	29.23	38.97	48.72	58.46	68.21	77.95	87.69
78	102	258	282	9.78	19.56	29.34	39.13	48.91	58.69	68.47	78.25	88.03
79	101	259	281	9.82	19.63	29.45	39.27	49.08	58.90	68.71	78.53	88.35
80	100	260	280	9.85	19.70	29.54	39.39	49.24	59.09	68.94	78.79	88.63
81	99	261	279	9.88	19.75	29.63	39.51	49.38	59.26	69.14	79.02	88.89
82	98	262	278	9.90	19.81	29.71	39.61	49.51	59.42	69.32	79.22	89.12
83	97	263	277	9.93	19.85	29.78	39.70	49.63	59.55	69.48	79.40	89.33
84	96	264	276	9.95	19.89	29.84	39.78	49.73	59.67	69.62	79.56	89.51
85	95	265	275	9.96	19.92	29.89	39.85	49.81	59.77	69.73	79.70	89.66
86	94	266	274	9.98	19.95	29.93	39.90	49.88	59.85	69.83	79.81	89.78
87	93	267	273	9.99	19.97	29.96	39.95	49.93	59.92	69.90	79.89	89.88
88	92	268	272	9.99	19.99	29.98	39.98	49.97	59.96	69.96	79.95	89.95
89	91	269	271	10.00	20.00	30.00	39.99	49.99	59.99	69.99	79.99	89.99
90	90	270	270	10.00	20.00	30.00	40.00	50.00	60.00	70.00	80.00	90.00

				Northings								
Bearing (degrees)				Distance (units of 10)								
+	−	−	+	1	2	3	4	5	6	7	8	9
00	180	180	360	10.00	20.00	30.00	40.00	50.00	60.00	70.00	80.00	90.00
01	179	181	359	10.00	20.00	30.00	39.99	49.99	59.99	69.99	79.99	89.99
02	178	182	358	9.99	19.99	29.98	39.98	49.97	69.96	79.95	79.95	89.95
03	177	183	357	9.99	19.97	29.96	39.95	49.93	59.92	69.90	79.89	89.88
04	176	184	356	9.98	19.95	29.93	39.90	49.88	59.85	69.83	79.81	89.78
05	175	185	355	9.96	19.12	29.89	39.85	49.81	59.77	69.73	79.70	89.66
06	174	186	354	9.95	19.89	29.84	39.78	49.73	59.67	69.62	79.56	89.51
07	173	187	353	9.93	19.85	29.78	39.70	49.63	59.55	69.48	79.40	89.33
08	172	188	352	9.90	19.81	29.71	39.61	49.51	59.42	69.32	79.22	89.12
09	171	189	351	9.88	19.75	29.63	39.51	49.38	59.26	69.14	79.02	88.89
10	170	190	350	9.85	19.70	29.54	39.39	49.24	59.09	68.94	98.79	88.63
11	169	191	349	9.82	19.63	29.45	39.27	49.08	58.90	68.71	78.53	88.35
12	168	192	348	9.78	19.56	29.34	39.13	48.91	58.69	68.47	78.25	88.03
13	167	193	347	9.74	19.49	29.23	38.97	48.72	58.46	68.21	77.95	87.69
14	166	194	346	9.70	19.41	29.11	38.81	48.51	58.22	67.92	77.62	87.33
15	165	195	345	9.66	19.32	28.98	38.64	48.30	57.96	67.61	77.27	86.93
16	164	196	344	9.61	19.23	28.84	38.45	48.06	57.68	67.29	76.90	86.51
17	163	197	343	9.56	19.13	28.69	38.25	47.82	57.38	66.94	76.50	86.07
18	162	198	342	9.51	19.02	28.53	38.04	47.55	57.06	66.57	76.08	85.60
19	161	199	341	9.46	18.91	28.37	37.82	47.28	56.73	66.19	75.64	85.10
20	160	200	340	9.40	18.79	28.19	37.59	46.98	56.38	65.78	75.18	84.57
21	159	201	339	9.34	18.67	28.01	37.34	46.68	56.01	65.35	74.69	84.02

TABLE 6—*continued*

Bearing (degrees)				Northings Distance (units of 10)								
+	−	−	+	1	2	3	4	5	6	7	8	9
22	158	202	338	9.27	18.54	27.82	37.09	46.36	55.63	64.90	74.17	83.45
23	157	203	337	9.21	18.41	27.62	36.82	46.03	55.23	64.44	73.64	82.85
24	156	204	336	9.14	18.27	27.41	36.54	45.68	54.81	63.95	73.08	82.22
25	155	205	335	9.06	18.13	27.19	36.25	45.32	54.38	63.44	72.50	81.57
26	154	206	334	8.99	17.98	26.96	35.95	44.94	53.93	62.92	71.90	80.89
27	153	207	333	8.91	17.82	26.73	35.64	44.55	53.46	62.37	71.28	80.19
28	152	208	332	8.83	17.66	26.49	35.32	44.15	52.98	61.81	70.64	79.47
29	151	209	331	8.75	17.49	26.24	34.98	43.73	52.48	61.22	69.97	78.72
30	150	210	330	8.66	17.32	25.98	34.64	43.30	51.96	60.62	69.28	77.94
31	149	211	329	8.57	17.14	25.72	34.29	42.86	51.43	60.00	68.57	77.15
32	148	212	328	8.48	16.96	25.44	33.92	42.40	50.88	59.36	67.84	76.32
33	147	213	327	8.39	16.77	25.16	33.55	41.93	50.32	58.71	67.09	75.48
34	146	214	326	8.29	16.58	24.87	33.16	41.45	49.74	58.03	66.32	74.61
35	145	215	325	8.19	16.38	24.57	32.77	40.96	49.15	57.34	65.53	73.72
36	144	216	324	8.09	16.18	24.27	32.36	40.45	48.54	56.63	64.72	72.81
37	143	217	323	7.99	15.97	23.96	31.95	39.93	47.92	55.90	63.89	71.88
38	142	218	322	7.88	15.76	23.64	31.52	39.40	47.28	55.16	63.04	70.92
39	141	219	321	7.77	15.54	23.31	31.09	38.86	46.63	54.40	62.17	69.94
40	140	220	320	7.66	15.32	22.98	30.64	38.30	45.96	53.62	61.28	68.94
41	139	221	319	7.55	15.09	22.64	30.19	37.74	45.28	52.83	60.38	67.92
42	138	222	318	7.43	14.86	22.29	29.73	37.16	44.59	52.02	59.45	66.88
43	137	223	317	7.31	14.63	21.94	29.25	36.57	43.88	51.19	58.51	65.82
44	136	224	316	7.19	14.39	21.58	28.77	35.97	43.16	50.35	57.55	64.74
45	135	225	315	7.07	14.14	21.21	28.28	35.36	42.43	49.50	56.57	63.64
46	134	226	314	6.95	13.89	20.84	27.79	34.73	41.68	48.63	55.57	62.52
47	133	227	313	6.82	13.64	20.46	27.28	34.10	40.92	47.74	54.56	61.38
48	132	228	312	6.69	13.38	20.07	26.77	33.46	40.15	46.84	53.53	60.22
49	131	229	311	6.56	13.12	19.68	26.24	32.80	39.36	45.92	52.48	59.05
50	130	230	310	6.43	12.86	19.28	25.71	32.14	38.57	45.00	51.42	57.85
51	129	231	309	6.29	12.59	18.88	25.17	31.47	37.76	44.05	50.35	56.64
52	128	232	308	6.16	12.31	18.47	24.63	30.78	36.94	43.10	49.25	55.41
53	127	233	307	6.02	12.04	18.05	24.07	30.09	36.11	42.13	48.15	54.16
54	126	234	306	5.88	11.76	17.63	23.51	29.39	35.27	41.14	47.02	52.90
55	125	235	305	5.74	11.47	17.21	22.94	28.68	34.41	40.15	45.89	51.62
56	124	236	304	5.59	11.18	16.78	22.37	27.96	33.55	39.14	44.74	50.33
57	123	237	303	5.45	10.89	16.34	21.79	27.23	32.68	38.12	43.57	49.02
58	122	238	302	5.30	10.60	15.90	21.20	26.50	31.80	37.09	42.39	47.69
59	121	239	301	5.15	10.30	15.45	20.60	25.75	30.90	36.05	41.20	46.35
60	120	240	300	5.00	10.00	15.00	20.00	25.00	30.00	35.00	40.00	45.00
61	119	241	299	4.85	9.70	14.54	19.39	24.24	29.09	33.94	38.78	43.63
62	118	242	298	4.69	9.39	14.08	18.78	23.47	28.17	32.86	37.56	42.25
63	117	243	297	4.54	9.08	13.62	18.16	22.70	27.24	31.78	36.32	40.86
64	116	244	296	4.38	8.77	13.15	17.53	21.92	26.30	30.69	35.07	39.45
65	115	245	295	4.23	8.45	12.68	16.90	21.13	25.36	29.58	33.81	38.04
66	114	246	294	4.07	8.13	12.20	16.27	20.34	24.40	28.47	32.54	36.61
67	113	247	293	3.91	7.81	11.72	15.63	19.54	23.44	27.35	31.26	35.17
68	112	248	292	3.75	7.49	11.24	14.98	18.73	22.48	26.22	29.97	33.71
69	111	249	291	3.58	7.17	10.75	14.33	17.92	21.50	25.09	28.67	32.25
70	110	250	290	3.42	6.84	10.26	13.68	17.10	20.52	23.94	27.36	30.78
71	109	251	289	3.26	6.51	9.77	13.02	16.28	19.53	22.79	26.05	29.30
72	108	252	288	3.09	6.18	9.27	12.36	15.45	18.54	21.63	24.72	27.81
73	107	253	287	2.92	5.85	8.77	11.69	14.62	17.54	20.47	23.39	26.31
74	106	254	286	2.76	5.51	8.27	11.03	13.78	16.54	19.29	22.05	24.81
75	105	255	285	2.59	5.18	7.76	10.35	12.94	15.53	18.12	20.71	23.29
76	104	256	284	2.42	4.84	7.26	9.68	12.10	14.52	16.93	19.35	21.77
77	103	257	283	2.25	4.50	6.75	9.00	11.25	13.50	15.75	18.00	20.25
78	102	258	282	2.08	4.16	6.24	8.32	10.40	12.48	14.55	16.63	18.71
79	101	259	281	1.91	3.82	5.72	7.63	9.54	11.45	13.36	15.26	17.18

TABLE 6—*continued*

Bearing (degrees)				Northings Distance (units of 10)								
+	−	−	+	1	2	3	4	5	6	7	8	9
80	100	260	280	1.74	3.47	5.21	6.95	8.68	10.42	12.16	13.89	15.63
81	99	261	279	1.56	3.13	4.69	6.26	7.82	9.39	10.95	12.51	14.08
82	98	262	278	1.39	2.78	4.18	5.57	6.96	8.35	9.74	11.13	12.53
83	97	263	277	1.22	2.44	3.66	4.88	6.09	7.31	8.53	9.75	10.97
84	96	264	276	1.05	2.09	3.14	4.18	5.23	6.27	7.32	8.36	9.41
85	95	265	275	0.87	1.74	2.61	3.49	4.36	5.23	6.10	6.97	7.84
86	94	266	274	0.70	1.40	2.09	2.79	3.49	4.19	4.88	5.58	6.28
87	93	267	273	0.52	1.05	1.57	2.09	2.62	3.14	3.66	4.19	4.71
88	92	268	272	0.35	0.70	1.05	1.40	1.75	2.09	2.44	2.79	3.14
89	91	269	271	0.17	0.35	0.52	0.70	0.87	1.05	1.22	1.40	1.57
90	90	270	270	0.00	0.00	0.00	0.00	0.00	0.00	0.00	0.00	0.00

Example of use of the table:

The line *AB* is 724.38 m in length on an azimuth of $236° 07' 56''$. To find approximate Eastings, read the line for $236°$.

Eastings for 700 m = 580.3
20 = 16.6
4 = 3.3
.4 = .3
—————
−600.5

Similarly Northings 700 m = 391.4
20 = 11.2
4 = 2.2
.4 = .2
—————
−405.0

Correct figures are   Eastings   −601.47
Northings −403.68

The approximate check can be improved by interpolating between degrees, but the table is intended only as an aid to rapid checking for gross error, and should not replace careful calculation. It is important to take out the correct + or − sign.

TABLE 7   Tilt correction for horizontal angles measured with the sextant

The table gives correction values for angles from $5°$ to $125°$ between elevations of $-10°$ and $+10°$ for two angles b_A and b_B. As these are interchangeable in the calculation only b_A is given in full, e.g. to find the correction for $b_A = +6°$ and $b_B = -2°$, use $b_A = -2°$ and $b_B = +6°$. Corrections for two negative angles are identical with those for the corresponding positive angles, e.g. for $b_A = -5°$ and $b_B = -2°$, use $b_A = +2°$ and $b_B = +5°$.

Three limiting lines are used in the table, for accuracies of $01'30''$, $03'30''$ and $08'30''$, corresponding approximately to a plottable error of $0·5$ m on a scale of $1:2000$ at ranges of 1000, 500 and 200 m respectively, for vertical angles measured accurately to the nearest $10'$ with an Abney clinometer. If the angles concerned fall outside the accuracy limits permitted, calculate in full from the formula

$$\cos Y = \frac{\cos X - \sin b_A \, \sin b_B}{\cos b_A \, \cos b_B}$$

where X is the observed angle and Y is the true horizontal angle, and/or use multiple angles to reduce the rate of change of the correction.

The extent to which corrections are calculated depends on scale, accuracy required and rate of change. With practice, corrections can be taken out by eye within acceptable limits, but an example is given of calculation in full.

Required to calculate the correction for $A\hat{O}B = 68°24'36''$ when $b_A = +02°20'$ and $b_B = +03°10'$.

From table
$65°\ b_A = 2°\ b_B = 3°$ correction $= 03'46''$ and $b_B = 4°$ correction $= 04'22''$
$70°\qquad\quad\ b_B = 3°$ correction $= 04'13''$ and $b_B = 4°$ correction $= 05'07''$

whence

$65°\ b_A = 2°\ b_B = 3°10'$ gives $\ 03'46'' + \frac{1}{6} . (04'22'' - 03'46'')$
$= 03'46'' + 6'' = 03'52''$

$70°\qquad\qquad\qquad\qquad$ gives $\ 04'13'' + \frac{1}{6} . (05'07'' - 04'13'')$
$= 04'13'' + 9'' = 04'22''$

When the observed angle increases from $65°$ to $70°$, the correction increases from $03'52''$ to $04'22''$.

When the observed angle increases from $65°$ to $68°24'36''$ the correction will increase from $03'52''$ to $04'12''$.

This is for $A\hat{O}B = 68°24'36''$ and $b_A = 2°$.

A similar calculation can be made for $b_A = 3°$.

For $A\hat{O}B = 68°24'36''$ and $b_A = 3°$ the correction is $06'45''$.

The change of b_A from $2°$ to $3°$ causes the correction to change from $04'12''$ to $06'45''$, i.e. a change of $02'33''$.

The change of b_A from $2°$ to $2°20'$ will cause the correction to change by $\frac{1}{3} . (02'33'')$, i.e. to change from $04'12''$ to $05'03''$.

The correction is +ve, so the true horizontal angle becomes $68°24'36'' + 05'03'' = 68°29'39''$.

TABLE 7—*continued*

b_A = $-10°$ b_B =	+1	+2	+3	+4	+5	+6	+7	+8	+9	+10
5°										
10										
15										
20										
25										
30										
35										
40										
45										
50	−59′11″									
55	50 41	−64′54″								
60	43 18	56 34	−70′29″							
65	36 45	49 15	62 17	−75′52″						
70	30 48	42 42	55 00	67 46	−80′57″					
75	25 20	36 45	48 29	60 31	72 54	−85′37″				
80	20 12	31 15	42 30	53 58	65 40	77 36	−89′46″			
85	15 19	26 05	36 58	47 58	59 05	70 21	81 44	−93′17″		
90	10 35	21 10	31 46	42 23	53 02	63 43	74 26	85 12	−96′02″	
95	05 56	16 25	26 49	37 09	47 24	57 36	67 44	77 48	87 50	−97′50″
100	−01 18	11 46	22 03	32 10	42 06	51 53	61 31	70 59	80 18	89 29
105	+03 24	07 08	17 23	27 22	37 04	46 31	55 42	64 38	73 19	81 46
110	08 14	−02 27	12 45	22 41	32 14	41 25	50 14	58 41	66 48	74 33
115	13 19	+02 22	08 05	18 03	27 31	36 31	45 02	53 05	60 40	67 47
120	18 44	07 25	−03 18	13 23	22 53	31 46	40 03	47 45	54 52	61 24
125	+24 39	+12 48	+01 42	−08 38	−18 14	−27 06	−35 15	−42 40	−49 21	−55 20

b_A = $-9°$ b_B =	+1	+2	+3	+4	+5	+6	+7	+8	+9	+10
5°										
10										
15										
20										
25										
30										
35										
40										
45	−57′18″									
50	49 06	−63′03″								
55	42 10	55 00	−68′40″							
60	36 08	48 07	60 46	−74′05″						
65	30 47	42 04	53 54	66 15	−79′10″					
70	25 56	36 41	47 49	59 24	71 24	−83′51″				
75	21 29	31 47	42 22	53 17	64 32	76 06	−88′00″			
80	17 18	27 15	37 25	47 47	58 22	69 10	80 13	−91′31″		
85	13 20	23 02	32 50	42 44	52 46	62 56	73 13	83 39	−94′14″	
90	09 30	19 00	28 32	38 04	47 38	57 14	66 52	76 32	86 15	−96′01″
95	05 45	15 09	24 28	33 43	42 53	51 59	61 02	70 00	78 57	87 50
100	−02 00	11 22	20 34	29 35	38 26	47 07	55 39	64 00	72 14	80 18
105	+01 47	07 38	16 46	25 38	34 14	42 33	50 38	58 27	66 00	73 19
110	05 41	03 52	13 01	21 48	30 13	38 15	45 55	53 14	60 12	66 48
115	09 45	−00 00	09 16	18 03	26 20	34 09	41 29	48 21	54 44	60 40
120	14 05	+04 00	05 27	14 18	22 33	30 12	37 15	43 43	49 35	54 52
125	+18 49	+08 17	−01 29	−10 31	−18 48	−26 22	−33 12	−39 18	−44 41	−49 21

TABLE 7—*continued*

$b_A =$ $b_B =$	$-8°$ $+1$	$+2$	$+3$	$+4$	$+5$	$+6$	$+7$	$+8$	$+9$	$+10$
5°										
10										
15										
20										
25										
30										
35										
40	−54′32″									
45	46 33	−60′18″								
50	40 00	52 29	−65′56″							
55	34 27	45 57	58 15	−71′22″						
60	29 37	40 21	51 43	63 45	−76′28″					
65	25 21	35 27	46 04	57 13	68 54	−81′09″				
70	21 29	31 04	41 04	51 29	62 19	73 35	−85′19″			
75	17 56	27 07	36 36	46 23	56 30	66 56	77 42	−88′48″		
80	14 37	23 29	32 32	41 48	51 17	60 59	70 56	81 06	−91′31″	
85	11 28	20 05	28 48	37 38	46 35	55 39	64 51	74 11	83 39	−93′17″
90	08 26	16 52	25 19	33 47	42 16	50 47	59 20	67 54	76 32	85 12
95	05 28	13 48	22 02	30 12	38 18	46 19	54 17	62 10	70 00	77 48
100	−02 31	10 48	18 54	26 50	34 36	42 11	49 37	56 54	64 00	70 59
105	+00 28	07 51	15 52	23 38	31 07	38 20	45 18	52 00	58 27	64 38
110	03 31	04 53	12 54	20 33	27 49	34 43	41 15	47 26	53 14	58 41
115	06 42	−01 52	09 57	17 33	24 40	31 18	37 27	43 08	48 21	53 05
120	10 05	+01 15	06 59	14 36	21 36	28 01	33 51	39 05	43 43	47 45
125	+13 45	+04 33	−03 55	−11 38	−18 37	−24 53	−30 25	−35 13	−39 18	−42 40

$b_A =$ $b_B =$	$-7°$ $+1$	$+2$	$+3$	$+4$	$+5$	$+6$	$+7$	$+8$	$+9$	$+10$
5°										
10										
15										
20										
25										
30										
35	−51′01″									
40	43 11	−56′57″								
45	36 58	49 08	−62′25″							
50	31 51	42 54	54 54	−67′51″						
55	27 31	37 42	48 39	60 24	−72′58″					
60	23 46	33 14	43 21	54 07	65 33	−77′40″				
65	20 27	29 21	38 46	48 43	59 12	70 14	−81′49″			
70	17 27	25 53	34 44	43 59	53 40	63 46	74 19	−85′19″		
75	14 42	22 46	31 08	39 48	48 47	58 06	67 44	77 42	−88′00″	
80	12 08	19 54	27 52	36 03	44 26	53 02	61 52	70 56	80 13	−89′4(
85	09 42	17 15	24 53	32 39	40 30	48 30	56 36	64 51	73 13	81 4
90	07 22	14 44	22 07	29 31	36 56	44 22	51 50	59 20	66 52	74 2(
95	05 05	12 21	19 32	26 37	33 38	40 35	47 28	54 17	61 02	67 4
100	02 50	10 02	17 04	23 55	30 35	37 06	43 26	49 37	55 39	61 3
105	−00 34	07 46	14 42	21 21	27 44	33 51	39 42	45 18	50 38	55 4
110	+01 45	05 31	12 24	18 55	25 03	30 49	36 13	41 15	45 55	50 1
115	04 09	03 14	10 09	16 34	22 30	27 58	32 57	37 27	41 29	45 0
120	06 42	−00 54	07 53	14 17	20 04	25 15	29 51	33 51	37 15	40 0
125	09 28	+01 34	05 36	12 01	17 43	22 40	26 54	30 25	33 12	35 1

TABLE 7—*continued*

$b_A =$	$-6°$									
$b_B =$	+1	+2	+3	+4	+5	+6	+7	+8	+9	+10
5°										
10										
15										
20										
25										
30	−46′53″									
35	39 07	−52′38″								
40	33 12	45 04	−58′16″							
45	28 30	39 07	50 50	−63′42″						
50	24 38	34 16	44 50	56 21	−68′49″					
55	21 22	30 14	39 52	50 16	61 29	−73′30″				
60	18 33	26 47	35 39	45 09	55 19	66 09	−77′40″			
65	16 03	23 46	32 00	40 45	50 02	59 51	70 14	−81′09″		
70	13 48	21 07	28 49	36 55	45 27	54 24	63 46	73 35	−83′51″	
75	11 45	18 43	25 59	33 32	41 24	49 35	58 06	66 56	76 06	−85′37″
80	09 50	16 32	23 25	30 30	37 48	45 19	53 02	60 59	69 10	77 36
85	08 02	14 31	21 05	27 46	34 34	41 28	48 30	55 39	62 52	70 21
90	06 18	12 37	18 56	25 16	31 37	37 59	44 22	50 47	57 14	53 43
95	04 38	10 49	16 56	22 58	28 55	34 47	40 35	46 19	51 59	57 36
100	02 58	09 05	15 03	20 49	26 25	31 50	37 06	42 11	47 07	51 53
105	−01 19	07 25	13 15	18 48	24 06	29 06	33 51	38 20	42 33	46 31
110	+00 22	05 46	11 32	16 54	21 55	26 33	30 49	34 43	38 15	41 25
115	02 07	04 06	09 51	15 05	19 52	24 09	27 58	31 18	34 09	36 31
120	03 57	02 25	08 11	13 21	17 55	21 53	25 15	28 01	30 12	31 46
125	+05 56	−00 40	−06 31	−11 40	−16 03	−19 44	−22 40	−24 53	−26 22	−27 06

$b_A =$	$-5°$									
$b_B =$	+1	+2	+3	+4	+5	+6	+7	+8	+9	+10
5°										
10										
15										
20										
25	−42′16″									
30	34 27	−47′59″								
35	28 50	40 25	−53′36″							
40	24 33	34 42	46 11	−59′01″						
45	21 09	30 13	40 24	51 42	−64′08″					
50	18 21	26 35	35 44	45 48	56 50	−68′49″				
55	16 00	23 33	31 51	40 56	50 49	61 29	−72′58″			
60	13 58	20 58	28 35	36 51	45 45	55 19	65 33	−76′28″		
65	12 10	18 43	25 46	33 20	41 25	50 02	59 12	68 54	−79′10″	
70	10 34	16 44	23 18	30 17	37 39	45 27	53 40	62 19	71 24	−80′57″
75	09 06	14 58	21 07	27 35	34 20	41 24	48 47	56 30	64 32	72 54
80	07 44	13 21	19 10	25 10	31 23	37 48	44 26	51 17	58 22	65 40
85	06 28	11 53	17 23	23 00	28 44	34 34	40 30	46 35	52 46	59 05
90	05 15	10 30	15 46	21 02	26 19	31 37	36 56	42 16	47 38	53 02
95	04 04	09 13	14 15	19 13	24 06	28 55	33 38	38 18	42 53	47 24
100	02 55	07 59	12 51	17 33	22 04	26 25	30 35	34 36	38 26	42 06
105	01 47	06 47	11 32	15 59	20 11	24 05	27 44	31 07	34 14	37 04
110	−00 37	05 38	10 16	14 31	18 24	21 55	25 03	27 49	30 13	32 14
115	+00 35	04 29	09 03	13 08	16 45	19 52	22 30	24 40	26 20	27 31
120	01 49	03 20	07 53	11 50	15 10	17 55	20 04	21 36	22 33	22 53
125	+03 09	−02 09	−06 44	−10 34	−13 41	−16 04	−17 43	−18 37	−18 48	−18 14

TABLE 7—*continued*

b_A =	−4°									
b_B =	+1	+2	+3	+4	+5	+6	+7	+8	+9	+10
5°										
10										
15										
20	−37'20"									
25	29 20	−42'57"								
30	23 59	35 17	−48'31"							
35	20 08	29 49	41 04	−53'55"						
40	17 13	25 41	35 28	46 34	−59'01"					
45	14 53	22 26	31 05	40 50	51 42	−63'42"				
50	12 59	19 49	27 33	36 13	45 48	56 21	−67'51"			
55	11 23	17 38	24 38	32 24	40 56	50 16	60 24	−71'22"		
60	10 00	15 46	22 10	29 11	36 51	45 09	54 07	63 45	−74'05"	
65	08 48	14 10	20 03	26 26	33 20	40 45	48 43	57 13	66 15	−75'52"
70	07 43	12 46	18 12	24 02	30 17	36 55	43 59	51 29	59 24	67 46
75	06 44	11 30	16 34	21 56	27 35	33 32	39 48	46 23	53 17	60 31
80	05 50	10 23	15 07	20 03	25 10	30 30	36 03	41 48	47 47	53 58
85	05 00	09 21	13 48	18 21	23 00	27 46	32 39	37 38	42 44	47 58
90	04 12	08 34	12 36	16 49	21 02	25 16	29 31	33 47	38 04	42 23
95	03 26	07 30	11 30	15 24	19 13	22 58	26 37	30 12	33 43	37 09
100	02 41	06 40	10 29	14 06	17 33	20 49	23 55	26 50	29 35	32 10
105	01 57	05 53	09 31	12 54	15 59	18 48	21 21	23 38	25 38	27 22
110	01 13	05 07	08 38	11 46	14 31	16 54	18 55	20 33	21 48	22 41
115	−00 28	04 22	07 47	10 42	13 08	15 06	16 34	17 33	18 03	18 03
120	+00 18	03 38	06 58	09 42	11 50	13 21	14 17	14 36	14 18	13 23
125	+01 07	−02 54	−06 11	−08 45	−10 34	−11 40	−12 01	−11 38	−10 31	−08 38

b_A =	−3°									
b_B =	+1	+2	+3	+4	+5	+6	+7	+8	+9	+10
5°										
10										
15	−32'17"									
20	23 49	−37'41"								
25	18 48	29 47	−43'10"							
30	15 26	24 32	35 33	−48'31"						
35	13 00	20 48	30 08	41 04	−53'36"					
40	11 10	17 58	26 04	35 28	46 11	−58'16"				
45	09 42	15 45	22 52	31 05	40 24	50 50	−62'25"			
50	08 31	13 57	20 18	27 33	35 44	44 50	54 54	−65'56"		
55	07 31	12 28	18 10	24 38	31 51	39 52	48 39	58 15	−68'40"	
60	06 40	11 13	16 23	22 10	28 35	35 39	43 21	51 43	60 46	−70'29"
65	05 55	10 07	14 50	20 03	25 46	32 00	38 46	46 04	53 54	62 17
70	05 15	09 11	13 30	18 12	23 18	28 49	34 44	41 04	47 49	55 00
75	04 40	08 21	12 19	16 34	21 07	25 59	31 08	36 36	42 22	48 29
80	04 07	07 35	11 15	15 07	19 10	23 25	27 52	32 32	37 25	42 30
85	03 37	06 54	10 18	13 48	17 23	21 05	24 53	28 48	32 50	36 58
90	03 09	06 17	09 27	12 36	15 46	18 56	22 07	25 19	28 32	31 46
95	02 42	05 43	08 39	11 30	14 15	16 56	19 32	22 02	24 28	26 49
100	02 16	05 11	07 55	10 29	12 51	15 03	17 04	18 54	20 34	22 03
105	01 51	04 41	07 14	09 31	11 32	13 15	14 42	15 52	16 46	17 23
110	01 26	04 13	06 37	08 38	10 16	11 32	12 24	12 54	13 01	12 45
115	01 02	03 46	06 00	07 47	09 03	09 51	10 09	09 57	09 16	08 05
120	00 36	03 20	05 27	06 58	07 53	08 11	07 53	06 59	05 27	−03 18
125	−00 10	−02 54	−04 55	−06 11	−06 44	−06 32	−05 36	−03 55	−01 29	+01 42

TABLE 7—*continued*

b_A =	$-2°$									
b_B =	+1	+2	+3	+4	+5	+6	+7	+8	+9	+10

5°

	+1	+2	+3	+4	+5	+6	+7	+8	+9	+10
10	−27′33″									
15	18 03	−32′25″								
20	13 24	24 00	−37′41″							
25	10 37	19 01	29 47	−42′57″						
30	08 45	15 42	24 32	35 17	−47′59″					
35	07 24	13 20	20 48	29 49	40 25	−52′38″				
40	06 23	11 32	17 58	25 41	34 42	45 04	−56′47″			
45	05 35	10 08	15 45	22 26	30 13	39 07	49 08	−60′18″		
50	04 56	09 05	13 57	19 49	26 35	34 16	42 54	52 29	−63′03″	
55	04 24	08 04	12 28	17 38	23 33	30 14	37 42	45 57	55 00	−64′54″
60	03 56	07 16	11 13	15 46	20 58	26 47	33 14	40 21	48 07	56 34
65	03 32	06 35	10 08	14 10	18 43	23 46	29 21	35 27	42 04	49 15
70	03 11	05 59	09 11	12 46	16 44	21 07	25 53	31 04	36 41	42 42
75	02 52	05 28	08 21	11 30	14 58	18 43	22 46	27 07	31 47	36 45
80	02 35	05 00	07 36	10 23	13 21	16 32	19 54	23 29	27 15	31 15
85	02 20	04 35	06 55	09 21	11 53	14 31	17 15	20 05	23 02	26 05
90	02 06	04 12	06 17	08 24	10 30	12 37	14 44	16 52	19 00	21 10
95	01 52	03 50	05 43	07 30	09 13	10 49	12 21	13 48	15 09	16 25
100	01 40	03 31	05 11	06 40	07 59	09 06	10 02	10 48	11 22	11 46
105	01 28	03 13	04 41	05 53	06 47	07 25	07 46	07 51	07 38	07 08
110	01 17	02 56	04 13	05 07	05 38	05 46	05 31	04 53	03 52	−02 27
115	01 05	02 40	03 46	04 22	04 29	04 06	03 14	−01 52	−00 00	+02 22
120	00 54	02 25	03 20	03 38	03 20	02 25	−00 54	+01 15	+04 00	07 25
125	−00 43	−02 11	−02 54	−02 54	−02 09	−00 40	+01 34	+04 33	+08 17	+12 48

b_A =	$-1°$									
b_B =	+1	+2	+3	+4	+5	+6	+7	+8	+9	+10

5°

	+1	+2	+3	+4	+5	+6	+7	+8	+9	+10
10	−12′06″									
15	07 59	−18′03″	−32′17″							
20	05 57	13 24	23 49	−37′20″						
25	04 44	10 37	18 48	29 20	−42′16″					
30	03 55	08 45	15 26	23 59	34 27	−46′53″				
35	03 19	07 24	13 00	20 08	28 50	39 07	−51′00″			
40	02 53	06 23	11 10	17 13	24 33	33 12	43 11	−54′32″		
45	02 32	05 35	09 42	14 53	21 09	28 30	36 58	46 33	−57′18″	
50	02 15	04 56	08 31	12 59	18 21	24 38	31 51	40 00	49 06	−59′11″
55	02 00	04 24	07 31	11 23	16 00	21 22	27 31	34 27	42 10	50 41
60	01 49	03 56	06 40	10 00	13 58	18 33	23 46	29 37	36 08	43 18
65	01 39	03 32	05 55	08 48	12 10	16 03	20 27	25 21	30 47	36 45
70	01 30	03 11	05 15	07 43	10 34	13 48	17 27	21 29	25 56	30 48
75	01 22	02 52	04 40	06 44	09 06	11 45	14 42	17 56	21 29	25 20
80	01 15	02 35	04 07	05 50	07 44	09 50	12 08	14 37	17 18	20 12
85	01 09	02 20	03 37	05 00	06 28	08 02	09 42	11 28	13 20	15 19
90	01 03	02 06	03 09	04 12	05 15	06 18	07 22	08 26	09 30	10 35
95	00 58	01 53	02 42	03 26	04 04	04 38	05 05	05 28	05 45	05 56
100	00 53	01 40	02 16	02 41	02 55	02 58	02 50	−02 31	−02 00	−01 18
105	00 48	01 28	01 48	01 51	01 57	−01 19	−00 34	+00 28	+01 47	+03 24
110	00 44	01 17	01 26	01 13	−00 37	+00 22	+01 45	03 31	05 41	08 14
115	00 40	01 05	01 02	−00 28	+00 35	02 07	04 09	06 42	09 45	13 19
120	00 36	00 54	00 36	+00 18	01 49	03 57	06 42	10 05	14 05	18 44
125	−00 33	−00 43	−00 10	+01 07	+03 09	+05 56	+09 28	+13 45	+18 49	+24 39

TABLE 7—*continued*

$b_A = 0$ $b_B = 0$	0	+1	+2	+3	+4	+5	+6	+7	+8	+9	+10
5°	nil	−06'03"									
10	nil	02 59	−12'00"								
15	nil	01 57	07 51	−17'47"							
20	nil	01 26	05 46	13 02	−23'17"						
25	nil	01 07	04 30	10 09	18 06	−28'25"					
30	nil	00 54	03 38	08 11	14 36	22 53	−33'04"	−45'14"			
35	nil	00 45	03 00	06 45	12 01	18 50	27 12	37 09	−48'45"		
40	nil	00 37	02 30	05 38	10 01	15 42	22 39	30 56	40 33	−51'32"	
45	nil	00 31	02 06	04 43	08 24	13 09	18 59	25 55	33 57	43 07	
50	nil	00 26	01 46	03 58	07 03	11 02	15 55	21 43	28 27	36 07	−44'45'
55	nil	00 22	01 28	03 18	05 53	09 12	13 17	18 07	23 43	30 06	37 17
60	nil	00 18	01 13	02 43	04 51	07 35	10 57	14 55	19 32	24 48	30 42
65	nil	00 15	00 59	02 12	03 55	06 08	08 50	12 03	15 46	20 00	24 46
70	nil	00 11	00 46	01 43	03 03	04 47	06 54	09 24	12 18	15 37	19 19
75	nil	00 08	00 34	01 16	02 15	03 31	05 04	06 55	09 03	11 29	14 13
80	nil	00 06	00 22	00 50	01 29	02 19	03 20	04 33	05 57	07 33	09 21
85	nil	−00 03	−00 11	−00 25	−00 44	−01 09	−01 39	−02 15	−02 57	−03 45	−04 38
90	nil	nil	nil	nil	nil	nil	nil	nil	nil	nil	nil
95	nil	+00 03	+00 11	+00 25	+00 44	+01 09	+01 39	+02 15	+02 57	+03 45	+04 38
100	nil	00 06	00 22	00 50	01 29	02 19	03 20	04 33	05 57	07 33	09 21
105	nil	00 08	00 34	01 16	02 15	03 31	05 04	06 55	09 03	11 29	14 13
110	nil	00 11	00 46	01 43	03 03	04 47	06 54	09 24	12 18	15 37	19 19
115	nil	00 15	00 59	02 12	03 55	06 08	08 50	12 03	15 46	20 00	24 46
120	nil	00 18	01 13	02 43	04 51	07 35	10 57	14 55	19 32	24 48	30 42
125	nil	+00 22	+01 28	+03 18	+05 53	+09 12	+13 17	+18 07	+23 43	+30 06	+37 17

$b_A = +1°$ $b_B = +1$	+2	+3	+4	+5	+6	+7	+8	+9	+10
5°	+00'03"	−05'57"							
10	00 05	02 48	−11'44"						
15	00 08	01 41	07 27	−17'14"					
20	00 11	01 04	05 13	12 18	−22'22"				
25	00 14	00 40	03 48	09 14	16 57	−27'03"			
30	00 17	00 21	02 48	07 04	13 12	21 13	−31'08"		
35	00 20	−00 05	02 00	05 26	10 22	16 52	24 54	−34'33"	
40	00 23	+00 08	01 21	04 06	08 07	13 25	20 00	27 55	−37'09"
45	00 26	00 20	00 48	02 59	06 14	10 34	15 58	22 28	30 04
50	00 29	00 32	−00 18	02 00	04 37	08 06	12 30	17 49	24 04
55	00 33	00 43	+00 10	01 07	03 09	05 56	09 28	13 45	18 49
60	00 36	00 54	00 36	−00 18	01 49	03 57	06 42	10 05	14 05
65	00 40	01 05	01 02	+00 28	−00 35	02 07	04 09	06 42	09 45
70	00 44	01 17	01 26	01 13	+00 37	−00 22	−01 45	03 31	05 41
75	00 48	01 28	01 51	01 57	01 47	+01 19	+00 34	−00 28	−01 47
80	00 53	01 40	02 16	02 41	02 55	02 58	02 50	+02 31	+02 00
85	00 58	01 52	02 42	03 26	04 04	04 38	05 05	05 28	05 45
90	01 03	02 06	03 09	04 12	05 15	06 18	07 22	08 26	09 30
95	01 09	02 20	03 37	05 00	06 28	08 02	09 42	11 28	13 20
100	01 15	02 35	04 07	05 50	07 44	09 50	12 08	14 37	17 18
105	01 22	02 52	04 40	06 44	09 06	11 45	14 42	17 56	21 29
110	01 30	03 11	05 15	07 43	10 34	13 48	17 27	21 29	25 56
115	01 39	03 32	05 55	08 48	12 10	16 03	20 27	25 21	30 47
120	01 49	03 56	06 40	10 00	13 58	18 33	23 46	29 37	36 08
125	02 00	04 24	07 31	11 23	16 00	21 22	27 31	34 27	42 10

Note: In the second table a first data column headed $b_B = +1$ precedes the +2 column; its values run 5° +00'03", 10 00 05, 15 00 08, 20 00 11, 25 00 14, 30 00 17, 35 00 20, 40 00 23, 45 00 26, 50 00 29, 55 00 33, 60 00 36, 65 00 40, 70 00 44, 75 00 48, 80 00 53, 85 00 58, 90 01 03, 95 01 09, 100 01 15, 105 01 22, 110 01 30, 115 01 39, 120 01 49, 125 02 00, and the +10 column continues beyond the page edge.

TABLE 7—*continued*

b_A = +2°								
b_B = +2	+3	+4	+5	+6	+7	+8	+9	+10
5° +00′11″	−05′47″							
10 00 22	02 26							
15 00 33	01 08	06 46	−16′25″					
20 00 44	−00 20	04 18	11 12	−21′06″				
25 00 55	+00 16	02 39	07 50	15 21	−25′13″			
30 01 07	00 47	01 23	05 23	11 15	18 59	−28′39″		
35 01 19	01 14	−00 21	03 27	08 04	14 14	21 57	−31′17″	
40 01 32	01 40	+00 33	01 49	05 27	10 22	16 35	24 07	−33′00″
45 01 44	02 05	01 23	−00 23	03 12	07 05	12 04	18 08	25 19
50 01 57	02 30	02 09	+00 56	−01 11	04 12	08 06	12 56	18 42
55 02 11	02 55	02 54	02 09	+00 40	−01 34	04 33	08 17	12 48
60 02 25	03 20	03 38	03 20	02 25	+00 54	−01 15	04 01	07 25
65 02 40	03 46	04 22	04 29	04 06	03 14	+01 52	+00 00	−02 22
70 02 56	04 13	05 07	05 38	05 46	05 31	04 53	03 52	+02 27
75 03 13	04 41	05 53	06 47	07 25	07 46	07 51	07 38	07 08
80 03 31	05 11	06 40	07 59	09 06	10 02	10 48	11 22	11 46
85 03 51	05 43	07 30	09 13	10 49	12 21	13 48	15 09	16 25
90 04 12	06 17	08 24	10 30	12 37	14 44	16 52	19 01	21 10
95 04 35	06 55	09 21	11 53	14 31	17 15	20 05	23 02	26 05
100 05 00	07 36	10 23	13 21	16 32	19 54	23 29	27 15	31 15
105 05 28	08 21	11 30	14 58	18 43	22 46	27 07	31 47	36 45
110 05 59	09 11	12 46	16 44	21 07	25 53	31 04	36 41	42 42
115 06 35	10 08	14 10	18 43	23 46	29 21	35 27	42 04	49 15
120 07 16	11 13	15 46	20 58	26 47	33 14	40 21	48 07	56 34
125 08 04	12 28	17 38	23 33	30 14	37 42	45 57	55 01	64 54

b_A = +3°							
b_B = +3	+4	+5	+6	+7	+8	+9	+10
5° +00′25″	−05′30″						
10 00 50	01 53	−10′39″					
15 01 15	−00 18	05 48	−15′20″				
20 01 40	+00 47	03 00	09 44	−19′28″			
25 02 06	01 40	−01 01	05 59	13 16	−22′55″		
30 02 32	02 28	+00 35	03 09	08 44	16 12	−25′35″	−36′58″
35 02 59	03 13	01 58	−00 48	05 06	10 56	18 20	27 21
40 03 26	03 57	03 14	01 14	−02 01	06 33	12 24	19 33
45 03 55	04 41	04 25	03 06	+00 43	−02 44	07 17	12 55
50 04 24	05 26	05 35	04 51	03 13	+00 42	−02 43	07 03
55 04 55	06 11	06 44	06 32	05 36	03 55	+01 29	−01 42
60 05 27	06 58	07 53	08 11	07 53	06 59	05 27	+03 18
65 06 00	07 47	09 03	09 51	10 09	09 57	09 16	08 05
70 06 37	08 38	10 16	11 32	12 24	12 54	13 01	12 45
75 07 15	09 31	11 32	13 15	14 42	15 52	16 46	17 23
80 07 55	10 29	12 51	15 03	17 04	18 54	20 34	22 03
85 08 39	11 30	14 15	16 56	19 32	22 02	24 28	26 49
90 09 27	12 36	15 46	18 56	22 07	25 19	28 32	31 46
95 10 18	13 48	17 23	21 05	24 53	28 48	32 50	36 58
100 11 15	15 07	19 10	23 25	27 52	32 32	37 25	42 30
105 12 19	16 34	21 07	25 59	31 08	36 36	42 22	48 29
110 13 30	18 12	23 18	28 49	34 44	41 04	47 49	55 00
115 14 50	20 03	25 46	32 00	38 46	46 04	53 54	62 17
120 16 23	22 10	28 35	35 39	43 21	51 43	60 46	70 29
125 +18 10	+24 38	+31 51	+39 52	+48 39	+58 15	+68 40	+79 56

TABLE 7—*continued*

b_A = +4° b_B =	+4	+5	+6	+7	+8	+9	+10
5°	+00′44″	−05′09″					
10	01 28	−01 09					
15	02 13	+00 48	04 34	−13′59″			
20	02 58	02 16	−01 21	07 54	−17′27″		
25	03 43	03 32	+01 04	03 40	10 44	−20′09″	
30	04 30	04 43	03 07	−00 20	05 38	12 50	−21′58″
35	05 18	05 52	04 57	+02 31	−01 27	06 58	14 03
40	06 07	07 01	06 40	05 04	+02 12	−01 58	07 26
45	06 57	08 10	08 21	07 28	05 31	+02 30	−01 37
50	07 50	09 21	10 00	09 46	08 38	06 36	+03 41
55	08 45	10 34	11 40	12 01	11 38	10 31	08 38
60	09 42	11 50	13 21	14 17	14 36	14 18	13 23
65	10 42	13 08	15 06	16 34	17 33	18 03	18 03
70	11 46	14 31	16 54	18 55	20 33	21 48	22 41
75	12 54	15 59	18 48	21 21	23 38	25 38	27 22
80	14 06	17 33	20 49	23 55	26 50	29 35	32 10
85	15 24	19 13	22 58	26 37	30 12	33 43	37 09
90	16 49	21 02	25 16	29 31	33 47	38 04	42 23
95	18 21	23 00	27 46	32 39	37 38	42 44	47 58
100	20 03	25 10	30 30	36 03	41 48	47 47	53 58
105	21 56	27 35	33 32	39 48	46 23	53 17	60 31
110	24 02	30 17	36 55	43 59	51 29	59 24	67 46
115	26 26	33 20	40 45	48 43	57 13	66 15	+75 52
120	29 11	36 50	45 09	54 07	63 45	+74 05	
125	+32 24	+40 56	+50 16	+60 24	+71 22		

b_A = +5° b_B =	+5	+6	+7	+8	+9	+10
5°	+01′09″	−04′42″				
10	02 18	−00 14	−08′51″			
15	03 27	+02 11	−03 03	−12′20″		
20	04 38	04 07	+00 42	05 41	−15′04″	
25	05 49	05 52	03 38	−00 52	07 43	−16′56″
30	07 02	07 32	06 13	+03 03	−01 59	08 54
35	08 17	09 12	08 36	06 30	+02 52	−02 19
40	09 34	10 51	10 54	09 41	07 11	+03 25
45	10 53	12 33	13 09	12 43	11 12	08 37
50	12 15	14 16	15 25	15 40	15 03	13 31
55	13 41	16 04	17 43	18 37	18 48	18 14
60	15 10	17 55	20 04	21 36	22 33	22 53
65	16 45	19 52	22 30	24 04	26 20	27 31
70	18 24	21 55	25 03	27 49	30 13	32 14
75	20 11	24 06	27 44	31 07	34 14	37 04
80	22 04	26 25	30 35	34 36	38 26	42 06
85	24 06	28 55	33 38	38 18	42 53	47 24
90	26 19	31 37	36 56	42 16	47 38	53 02
95	28 44	34 34	40 30	46 35	52 46	59 05
100	31 23	37 48	44 26	51 17	58 22	65 40
105	34 20	41 24	48 47	56 30	64 32	72 54
110	37 39	45 27	53 40	62 19	71 24	+80 57
115	41 25	50 02	59 12	68 54	+79 10	
120	45 45	55 19	65 33	+76 28		
125	+50 49	+61 29	+72 58			

TABLE 7—*continued*

	b_A = +6° b_B = +6	+7	+8	+9	+10	b_A = +7° b_B = +7	+8	+9	+10
5°	+01'39"	−04'09"				+02'15"	−03'31"		
10	03 19	+00 52	−07'40"			04 31	+02 10	−06'18"	
15	04 59	03 51	−01 15	−10'25"		06 48	05 48	+00 50	−08'14"
20	06 41	06 21	+03 06	−03 05	−12'19"	09 06	08 58	05 54	−00 07
25	08 24	08 40	06 41	+02 24	−04 13	11 27	11 58	10 12	+06 09
30	10 09	10 56	09 54	07 00	+02 16	13 50	14 55	14 10	11 34
35	11 57	13 12	12 56	11 10	07 52	16 17	17 53	17 58	16 32
40	13 47	15 28	15 54	15 04	12 58	18 48	20 53	21 42	21 16
45	15 42	17 48	18 51	18 51	17 48	21 24	23 57	25 28	25 55
50	17 40	20 12	21 50	22 36	22 28	24 06	27 08	29 17	30 33
55	19 44	22 40	24 53	26 22	27 06	26 54	30 25	33 12	35 15
60	21 53	25 15	28 01	30 12	31 46	29 51	33 51	37 15	40 03
65	24 09	27 58	31 18	34 09	36 31	32 57	37 27	41 29	45 02
70	26 33	30 49	34 43	38 15	41 25	36 13	41 15	45 55	50 14
75	29 06	33 51	38 20	42 33	46 31	39 42	45 18	50 38	55 42
80	31 50	37 06	42 11	47 07	51 53	43 26	49 37	55 39	61 31
85	34 47	40 35	46 19	51 59	57 36	47 28	54 17	61 02	67 44
90	37 59	44 22	50 47	57 14	63 43	51 50	59 20	66 52	74 26
95	41 28	48 30	55 39	62 56	70 21	56 36	64 51	73 13	81 44
100	45 19	53 02	60 59	69 10	77 36	61 52	70 56	80 13	+89 46
105	49 35	58 06	66 56	76 06	+85 37	67 44	77 42	+88 00	
110	54 24	63 46	73 35	+83 51		74 19	+85 19		
115	59 51	70 14	+81 09			+81 49			
120	66 09	+77 40							
125	+73 30								

	b_A = +8° b_B = +8	+9	+10	b_A = +9° b_B = +9	+10	b_A = +10° b_B = +10
5°	+02'57"	−02'47"		+03'45"	−01'58"	+04'38"
10	05 55	+03 39	−04'45"	07 30	+05 19	09 17
15	08 54	08 02	+03 12	11 17	10 34	13 58
20	11 55	11 58	09 05	15 07	15 21	18 42
25	14 59	15 44	14 13	19 00	20 00	23 31
30	18 07	19 29	19 00	22 58	24 38	28 26
35	21 19	23 15	23 41	27 02	−29 19	33 28
40	24 37	27 05	28 19	31 13	34 06	38 39
45	28 00	31 03	32 59	35 32	39 00	44 00
50	31 33	35 05	37 45	40 01	44 05	49 33
55	35 13	39 18	42 40	44 41	49 21	55 20
60	39 04	43 43	47 45	49 35	54 52	61 24
65	43 08	48 21	53 05	54 44	60 40	67 47
70	47 26	53 14	58 41	60 12	66 48	74 33
75	52 00	58 27	64 38	66 00	73 19	81 46
80	56 54	64 00	70 59	72 14	80 18	89 29
85	62 10	70 00	77 48	78 57	87 50	+97 50
90	67 54	76 32	85 12	86 15	+96 01	
95	74 11	83 39	+93 17	+94 14		
100	81 06	+91 31				
105	+88 48					
110						
115						
120						
125						

TABLE 8   Conversion constants

LENGTH				
1 mile	= 1.609 3480	km	logarithm	0.206 6500
	= 8	furlongs		0.903 0900
	= 1760	yards		3.245 5127
	= 5280	feet		3.722 6339
	= 63360	inches		4.801 8152
1 furlong	= 0.201 1685	km		1̄.303 5600
	= 10	chains		
	= 220	yards		2.342 4227
1 chain	= 66	feet		1.819 5439
	= 20.116 8494	m		1.303 5600
1 fathom	= 6	feet		0.778 1513
	= 1.828 8045	m		0.262 1673
1 yard	= 0.914 4022	m		1̄.961 1373
1 foot	= 0.304 8007	m		1̄.484 0160
1 inch	= 25.400 0625	mm		1.404 8348
1 nautical mile	= 6080	feet		3.783 9036
	= 1853.188 556	m		3.267 9196
1 cable	= 100	fathoms		
1 kilometre	= 0.621 3700	miles		1̄.793 3500
	= 1093.61061	yards		3.038 8627
1 metre	= 1.093 6106	yards		0.038 8627
	= 3.280 8318	feet		0.515 9840
	= 39.369 9820	inches		1.595 1652
1 mm	= 0.039 3700	inches		2̄.595 1652

AREA				
1 sq. mile	= 2.590 0010	sq. km	logarithm	0.413 2999
1 sq. yard	= 0.836 1314	sq. m		1̄.922 2745
1 sq. foot	= 0.092 9035	sq. m		2̄.968 0320
1 sq. inch	= 645.163 175	sq. mm		2.809 6696
1 acre	= 10	sq. chains		
	= 43560	sq. ft		4.639 0879
	= 0.404 6876	hectares		1̄.607 1199
1 sq. km	= 0.386 1003	sq. miles		1̄.586 7001
	= 1 195 984.168	sq. yds		6.077 7254
1 sq. m	= 1.195 9842	sq. yds		0.077 7254
	= 10.763 8575	sq. ft		1̄.031 9679
1 sq. cm	= 0.154 9995	sq. ins		1̄.190 3304
1 hectare	= 2.471 0418	acres		0.392 8801
	= 11959.8417	sq. yds		4.077 7254
	= 107 638.5751	sq. ft		5.031 9679

These figures for length and area are based on the British O.S.
Standard: 1 metre = 3.280 83183 feet 0.515 9840
The U.S.A. Standard is   1 metre = 3.280 83333 feet 0.515 9842
The Canadian Standard is   1 metre = 3.280 83990 feet 0.515 9851

Index

Abney clinometer, 54, 56, 103, 153-6
alidade, 160
alidade, telescopic, 85-8, 95, 160
altitudes, observed, 82
    resected, 85, 94
aneroid barometer, 111-17, 156-7
arrows, 50, 53
astronomical triangle, 2-3, 9, 173-4, 181
astronomy, xviii, 169
azimuths, 2-14, 36, 172-4, 182-5

base extension nets, 20
base lines, 17
Beaman arc, 86-8
bearings, forward and back, 59-61
    magnetic, 97, 166
bench marks, 43, 111
Bessel's method of resection, 90, 191-3
Bowditch's Rule, 63, 67, 100, 104, 127
braced quadrilaterals, 26-34
bubble displacement, 8-9

calculators, 166
centre-point figures, 36-8
chain, 52-3
chaining, 49-56, 62
checks, xviii, 46, 49, 65, 67, 126-7
collimation, horizontal, 144
    vertical, 123-5, 142-3, 160-1
compass, Brunton, 107-10
    prismatic, 97-107, 163-6
    trough, 91, 162

condition equations, 27-38
contouring, 94-6, 126
curvature of earth, 42-3, 45

danger circle (resection), 93
declination, 3, 11, 169-72, 174, 178

Eastings and Northings, 25, 31-5, 63-5
electronic distance measurement, xvii, 17
Equation of Time, 12, 175-6, 184
Ewing stadi-altimeter, 73-6
ex-meridian altitudes of sun, 2, 177, 184

Face Left (Face Right), 138
formlines, 56, 66, 106-7

Greenwich Mean Time, 11, 175-6, 184

heights, levelled, 43-6
    theodolite traverse, 66-7
    trigonometrical, 41-3
hour angle, 171-4

Indian clinometer, 81-3, 94, 161-2

latitude observations, 176-88
level, quickset, 119-22
    adjustment, 123-5
    use in contouring, 94-6
levelling, 44, 125-8
levelling staves, 121-2

local apparent time, 175, 184
local attraction (magnetic), 98
local mean time, 12, 175-6, 184
longitude, correction for, 12, 176
    observations for, 176-88

magnetic declination, 97
Magnetic North, 14, 97
micrometers, 70-1, 139-41

Nautical Almanac, 3
North line, 14

offsets, 54-5, 102
orientation, 1-2

parallax, instrumental, 120
    sextant, 77-8, 149-50
    solar, 10
plane table, 80-8, 89-96, 158-63
plottable error, xvi, 1, 20, 83
plotting of fixed points, 35, 66, 91
polar distance, 11-12, 173-4
position lines, 179-88
    plotting, 186-8

radio telephones, 167-8
ranging rods, 50-1
Reference Objects, 2, 79
refraction, 10-11, 42-3, 45
resection, partial, 89
    plane table, 89, 191-3
    trigonometrical, 80, 189-91
Right Ascension, 169-71, 180

satellite stations, 38-41
sextant, 76-80, 145-53
slant range, 69

solar time, 175
spring balance, 18, 50
stadia readings, 45, 68-76, 85-8, 94-6,
    125-6
Star Almanac, 4
station marks, xviii, 16, 48
station records, 17, 48
subtense measurements, 22-3, 117
sun azimuths, 2-14

tacheometry, 46, 67-76, 85-8, 125-6
tangent screws, 120, 130, 135-6
tape, linen, 50, 55, 95, 163
    steel, 18-19, 49-51
theodolite, 128-45
    angles, 56-9
    astronomical observation, 4-9
    detail fixation, 96
    levelling of, 133-4
    traverse calculation, 59-66
time, 174
    calculation of, 11-12
    observation for, 4-6
traverses, aneroid, 111-17
    compass, 97-107
    plane table, 80-8
    theodolite, xix, 47-67
triangle of error (resection), 92
triangulation, xix, 15-22
    adjustment, 24-35
trigonometrical functions, 192-8
tripod, target, 15

verniers, 155

well-conditioned angles, 40

zenith distance, 173-4